PARCC Math Prep

Grade 8

The Ultimate Step-by-Step Guide Plus

Two Full-Length PARCC Practice Tests

SCAN ME

Michael Smith

www.mathnotion.com

PARCC Math Prep Grade 8

Published By: The Math Notion

Web: www.mathnotion.com

Email: info@mathnotion.com

ISBN: 978-1-63620-204-4

The Math Notion

Michael Smith has been a math instructor for over a decade now. He launched the Math Notion. Since 2006, we have devoted our time to both teaching and developing exceptional math learning materials. As a test prep company, we have worked with thousands of students. We have used the feedback of our students to develop a unique study program that can be used by students to drastically improve their math scores fast and effectively. We have more than a thousand Math learning books including:

- **ACT Math Prep**
- **SAT Math Prep**
- **PSAT Math Prep**
- **Accuplacer Math Prep**
- **Common Core Math Prep**
- **many Math Education Workbooks, Study Guides, Practice and Exercise Books**

As an experienced Math test preparation company, we have helped many students raise their standardized test scores—and attend the colleges of their dreams: We tutor online and in person, we teach students in large groups, and we provide training materials and textbooks through our website and through Amazon.

You can contact us via email at:

info@mathnotion.com

How to Achieve a Perfect Score on the 8th grade PARCC Math Test?

The math preparation book covers all mathematics topics that will be the key to succeeding on the 8th grade PARCC math test. The step-by-step guide and hundreds of examples in this book can help you hone your math skills, boost your confidence, and be well prepared for the PARCC test. This new PARCC grade 8 self-teaching book offers extensive preparation and brush-up in math for those test-takers who plan to take the PARCC Test.

Updated the new PARCC prep for 2021, 2022, and beyond by top test prep experts. **Inside the PARCC math prep book, You'll Find:**

- Comprehensive 8th grade math review of the key concepts.
- Content 100% aligned with the latest PARCC math exam grade 8.
- Over 2,000 practice questions to help you master each math topic.
- Focus on the most challenging part of the PARCC 8th grade math test!
- 2 full-length practice exams (featuring new question types) with detailed answers

After completing the 8th grade math prep book, you will discover your strengths and weaknesses, gain confidence and a strong foundation, to succeed on the PARCC test. We have helped hundreds of thousands of people pass the PARCC test and achieve their education and career goals. Get math prep grade 8 for the PARCC review that you need to ace your exam.

It is an excellent investment in your future!

www.MathNotion.com

... So Much More Online!

- ✓ FREE Math Lessons

- ✓ More Math Learning Books!

- ✓ Mathematics Worksheets

- ✓ Online Math Tutors

- ✓ For a PDF Version of This Book

Please Visit www.mathnotion.com

Contents

Chapter 1 : Whole Numbers, Real Numbers, and Integers

Topics that you'll learn in this chapter:

- ➢ Rounding and Estimates

- ➢ Addition, Subtraction, Multiplication and Division Whole Number and Integers

- ➢ Arrange and ordering Integers and Numbers

- ➢ Comparing Integers, Order of Operations

- ➢ Mixed Integer Computations

- ➢ Integers and Absolute Value

"If people do not believe that mathematics is simple, it is only because they do not realize how complicated life is." — John von Neumann

Name: ....

Rounding

Rounding is replacing a number up or down to the closest number or the closest hundred, etc.

✓ First, you have to know the place value you'll round to.

✓ Second, you have to find the digit to the right of the place value you're rounding to. If it is 5 or greater, add 1 to the place value you're rounding to and put zero for all digits on its right side. If the digit to the right of the place value is smaller than 5 then keep the place value and put zero for all digits to the right.

EXAMPLE:

Round 64 to the closest ten.

The place value of ten is 6. The digit on the right side is 4 (which is smaller than 5). Now keep 6 and put zero for the digit on the right side. Now our answer is 60. 64 is rounded to the closest ten is 60, because 64 is closer to 60 than to 70.

PRACTICES:

Round each number to the underlined place value.

1) 88	2) 8.15
3) 4,315	4) 565
5) 1.331	6) 14.23
7) 2.429	8) 4.313
9) 2.997	10) 7.38

Score: ....

Answer Key	
1) 90	2) 8.0
3) 4,000	4) 570
5) 1.3	6) 14.2
7) 2.0	8) 4.31
9) 3.0	10) 7.0

Name: ....

Estimates

Estimating is a math policy used for approximating a number. To estimate *means* to make an irregular guess or calculation. To round means to make easier a known number by scaling it a little bit up or down.

- ✓ To estimate a math problem, round the numbers.
- ✓ For 2-digit numbers, you can usually round to the nearest tens, for 3-digit numbers, round to nearest hundreds, etc.
- ✓ Find the answer.

EXAMPLE:

Estimate the sum by rounding every number to the closest hundred. **153 + 426 =?**

153 is rounded to the closest hundred which is 200. Now 426 is rounded to the closest hundred which is 400.

Then: 200 + 400 = 600

PRACTICES:

Estimate the sum by rounding each added to the nearest ten.

1) 17 + 18	2) 94 + 81
3) 203 + 56	4) 55 + 33
5) 96 + 49	6) 99 + 324
7) 823 + 488	8) 466 + 276
9) 5,112 + 5,792	10) 1,245 + 2,459

Score: ....

Answer Key	
1) 40	2) 200
3) 260	4) 90
5) 150	6) 400
7) 1,300	8) 800
9) 11,000	10) 3,000

Name: ....

Whole Number Addition and Subtraction

- ✓ Arrange the numbers in line.
- ✓ Start with the unit place. (Ones place)
- ✓ Regroup if needed.
- ✓ Add or subtract the tens place.
- ✓ Continue with further digits.

EXAMPLE:

Find the sum. $285 + 145 =$?

First line up the numbers: $\begin{array}{r}285\\+145\\\hline\end{array}$ → Start with the unit place. (ones place) $5 + 5 = 10$,

Write 0 for ones place and keep 1, $\begin{array}{r}1\\285\\+145\\\hline 0\end{array}$, Add the tens place and the digit 1 we kept:

$1 + 8 + 4 = 13$, Write 3 and keep 1, $\begin{array}{r}1\,1\\285\\+145\\\hline 30\end{array}$

Continue with further digits → $1 + 2 + 1 = 4$ → $\begin{array}{r}1\,1\\285\\+145\\\hline 430\end{array}$

Find the difference. $976 - 453 =$?

First line up the numbers: $\begin{array}{r}976\\-453\\\hline\end{array}$, → Start with the unit place. $6 - 3 = 3$, $\begin{array}{r}976\\-453\\\hline 3\end{array}$,

Subtract the tens place. $7 - 5 = 2$, $\begin{array}{r}976\\-453\\\hline 23\end{array}$, Continue with further digits → $9 - 4 = 5$,

$\begin{array}{r}976\\-453\\\hline 523\end{array}$

PRACTICES:

Find the missing number.

1) 540 − .... = 100	2) 800 − .... = 220
3) .... − 2,650 = 6,700	4) 85,000 − 42,000 = ...
5) 1,280− .... = 420	6) 5,000 + 8,450 = ...
7) .... − 3,870 = 9,630	8) 12,310 − .... = 8,540

Solve.

9) A school had 708 students last year. If all last year students and 218 new students have registered for this year, how many students will there be in total?

10) Lisa had $856 dollars in her saving account. She gave $295 dollars to her brother, Tom. How much money does she have left?

Score: ....

Answer Key	
1) 440	2) 580
3) 9,350	4) 43,000
5) 860	6) 13,450
7) 13,500	8) 3,770
9) 926	10) 561

Name: ....

Whole Number Multiplication

- ✓ First you have to learn the times tables! To solve multiplication problems quick, you need to learn the times table. For example, 3 times 8 is 24 or 8 times 7 is 56.
- ✓ For multiplication, line up the numbers that you are multiplying.
- ✓ Start with the ones place and regroup if needed.
- ✓ Continue with further digits.

EXAMPLE:

Solve. $500 \times 30 = ?$

Line up the numbers: $\begin{array}{r} 500 \\ \times 30 \\ \hline \end{array}$, start with the ones place → $0 \times 500 = 0$, $\begin{array}{r} 500 \\ \times 30 \\ \hline 0 \end{array}$, Continue

with further digit which is 3. → $3 \times 500 = 1,500$, $\begin{array}{r} 500 \\ \times 30 \\ \hline 15,000 \end{array}$

PRACTICES:

Multiply the Number.

1) $120 \times 6 =$ _____	2) $160 \times 30 =$ _____
3) $600 \times 30 =$ _____	4) $420 \times 20 =$ _____
5) $250 \times 40 =$ _____	6) $600 \times 40 =$ _____
7) $215 \times 70 =$ _____	8) $540 \times 11 =$ _____
9) $121 \times 10 =$ _____	10) $254 \times 16 =$ _____

Score: ....

Answer Key	
1) 720	2) 4,800
3) 18,000	4) 8,400
5) 10,000	6) 24,000
7) 15,050	8) 5,940
9) 1,210	10) 4,064

Name: ....

Whole Number Division

- ✓ Division: A typical division problem: Dividend ÷ Divisor = Quotient
- ✓ In division, we want to find how many times a divisor is contained in a dividend. The result we obtain in a division problem is called quotient.
- ✓ First, the problem is written in division format. (Dividend is inside; divisor is outside)

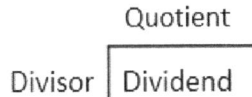

$$\text{Divisor}\,\overline{\left)\,\text{Dividend}\right.}^{\text{Quotient}}$$

EXAMPLE:

Solve. **234 ÷ 4 = ?**

First, write the problem in division format. $4\,\overline{\left)\,234\right.}$

Start from left digit of the dividend. 4 won't divide 2.

So, we have to choose another digit of the dividend. It is 3.

Now, we will find how many times 4 goes into 23 and the answer is 5.

Write 5 above the dividend part. 4 times 5 is 20. $4\,\overline{\left)\,234\right.}^{5}$

Write 20 below 23 and subtract. We get the answer 3.

Now take down the next digit which is 4 and find how many times 4 goes into 34?

The answer is 8. Write 8 above dividend.

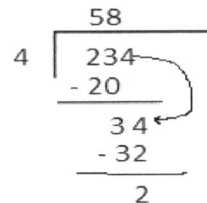

This is last step since there is no further digit left.

of the dividend to bring down.

The final answer is 58 and we have the remainder 2.

$$
\begin{array}{r}
58 \\
4\,\overline{\smash{\big)}\,234} \\
-20 \\
\hline
34 \\
-32 \\
\hline
2
\end{array}
$$

PRACTICES:

Divide the Number.

1) 450 ÷ 5 = _____	2) 320 ÷ 8 = _____
3) 125 ÷ 25 = _____	4) 720 ÷ 12 = _____
5) 588 ÷ 14 = _____	6) 299 ÷ 13 = _____
7) 869 ÷ 11 = _____	8) 801 ÷ 9 = _____
9) 493 ÷ 17 = _____	10) 600 ÷ 24 = _____

Score: ....

Answer Key	
1) −3	2) 12
3) 34	4) 12
5) −30	6) −27
7) 24	8) −2
9) 30	10) 27

Name: ....

Multiplying and Dividing Integers

- ✓ (positive) × (positive) = positive
- ✓ (positive) ÷ (positive) = positive
- ✓ (negative) × (negative) = positive
- ✓ (negative) ÷ (negative) = positive
- ✓ (negative) × (positive) = negative
- ✓ (negative) ÷ (positive) = negative
- ✓ (positive) × (negative) = negative
- ✓ (positive) ÷ (negative) = negative

÷ ╲ ×	+	−
+	+	−
−	−	+

EXAMPLE:

$(+5) × (+3) = 5 + 5 + 5 = 15$

The basic idea of multiplication is recurrent addition. Example: $5 × 3 = 5 + 5 + 5 = 15$

We know that division is the inverse operation of multiplication. So, $15 ÷ 3 = 5$ because $5 × 3 = 15$ In words, this expression says that 15 may be divided into 3 groups of 5 every because adding five thrice gives 15.

Divide (-91) by (-7)?

Examples on division of integers on different kinds of problems on integers are mentioned here step by step. $(-91) ÷ (-7) = 13$

PRACTICES:

Find each product and each quotient.

1) $(-8) × (-5)$	2) $72 ÷ 9$
3) $4 × (-5) × (-6)$	4) $(-95) ÷ (-5)$
5) $32 × (-4)$	6) $(-99) ÷ (-11)$
7) $(-12) × (-4)$	8) $(-123) ÷ 1$
9) $(-4) × (-3) × 5$	10) $(-0) ÷ 15$

Score: ....

Answer Key	
1) 40	2) 8
3) 120	4) 19
5) −128	6) 9
7) 48	8) −123
9) 60	10) 0

Name: ....

Arrange, Order, and Comparing Integers

- ✓ When we use a number line, numbers are increased when you go to the right.
- ✓ To compare numbers, you can use number line! As you go from left to right on the number line, you will find a greater number!
- ✓ Order integers from smallest to greatest.

EXAMPLE:

Order integers from to greatest.

$$(-11, -13, 7, -2, 12)$$

To compare numbers, you can use number line! When you see from left to right on the number line, you find a greater number!

$$-13 < -11 < -2 < 7 < 12$$

PRACTICES:

Order each set of integers from least to greatest.	Order each set of integers from greatest to least
1) $2, 6, -15, -11, 1$	2) $1, 17, 6, 8, 65, 2$
3) $9, -8, 3, -2, 11$	4) $-12, 6, -7, 2, -11$
5) $36, -12, 5, 1, -2$	6) $-14, 17, 7, 37, 9$
7) $31, 18, 0, -54, 9, -5$	8) $-54, 0, 14, 19, 15$
9) $-15, -25, -37, 7, 0, 9$	10) $12, 7, -1, -11, 9, -3$

Score: ....

Answer Key	
1) $-15, -11, 1, 2, 6$	2) $65, 17, 8, 6, 2, 1$
3) $-8, -2, 3, 9, 11$	4) $6, 2, -7, -11, -12$
5) $-12, -2, 1, 5, 36$	6) $37, 17, 9, 7, -14$
7) $-54, -5, 0, 9, 18, 31$	8) $19, 15, 14, 0, -54$
9) $-37, -25, -15, 0, 7, 9$	10) $12, 9, 7, -1, -3, -11$

Name: ....

Compare Integer

✓ If you want to compare numbers, you can use a number line! As you move from left to right on the number line, you will find a greater number!

EXAMPLE:

-5 ____ -1

When we compare two integers, we use the symbols $<$ and $>$.

$-5 < -1$ means that -5 is less than -1

PRACTICES:

Compare. Use >, =, <

1) 4 ____ 3	2) -22 ___ -11
3) 0 ____ -31	4) -41 ____ -12
5) -64 ____ 64	6) -142 ____ -148
7) 68 ____ 100	8) $(-15) \times 6$ ____ $5 \times (-18)$
9) 16 ____ $-(-16)$	10) 405 ____ -405

Score: ....

Answer Key	
1) >	2) <
3) >	4) <
5) <	6) >
7) <	8) =
9) =	10) >

Name: ....

Order of Operations

When you find more than one math operation, use PEMDAS:
- ✓ Parentheses
- ✓ Exponents
- ✓ Multiplication and Division (from left to right)
- ✓ Addition and Subtraction (from left to right)

EXAMPLE:

Solve. $(11 \times 5) - (12 - 25) =$

First you have to simplify inside parentheses: $(11 \times 5) - (12 - 25) = (55) - (-13) =$

Then: $55 + 13 = 68$

PRACTICES:

Evaluate each expression.

1) $24 - (8 \times 6)$	2) $5 \times 6 - (\frac{15}{11 - (-4)})$
3) $12 - (6 \times (-3))$	4) $(6 \times 7) + (-7)$
5) $(\frac{(-1)+4}{(-1)+(-2)}) \times (-9)$	6) $\frac{30}{2(9-(-1))-10}$
7) $58 - (6 \times 9)$	8) $13 + (4 \times 2)$
9) $((-3) + 15) \div (-3)$	10) $[(-8 \div 2) \div (2-4))$

Score: ....

Answer Key		
1) −24		2) 29
3) 30		4) 35
5) 9		6) 3
7) 4		8) 21
9) −4		10) 2

Name: ....

Integers and Absolute Value

✓ To find a definite value of a number, simply find its distance from 0 on number line! For example, the distance of 13 and −13 from zero on number line is 13!

EXAMPLE:

Solve. $\dfrac{|-18|}{9} \times |5-8| =$

First find $|-18|$, →the definite value of −18 is 18, then: $|-18| = 18$

$\dfrac{18}{9} \times |5-8| =$

Next, we solve $|5-8|$, → $|5-8| = |-3|$, the definite value of −3 is 3. $|-3| = 3$

Then: $\dfrac{18}{9} \times 3 = 2 \times 3 = 6$

PRACTICES:

Write absolute value of each number.

1) 62	2) − 32
3) − 11	4) 5

Evaluate.

5) $|-12| - |3| + 2$	6) $19 + |-5-14| - |2|$
7) $|-11| + |-9|$	8) $|91| - |-18| - 18$
9) $|-10+4| \times \dfrac{|-7\times5|}{7}$	10) $\dfrac{|-16\times3|}{2} \times |-12|$

Score: ....

Answer Key	
1) 62	2) 32
3) 11	4) 5
5) 11	6) 36
7) 20	8) 55
9) 30	10) 288

Chapter 2 : Fractions and Decimals

Topics that you'll learn in this chapter:

- ➢ Simplifying Fractions

- ➢ Adding and Subtracting Fractions, Mixed Numbers and Decimals

- ➢ Multiplying and Dividing Fractions, Mixed Numbers and Decimals

- ➢ Comparing and Rounding Decimals

- ➢ Converting Between Fractions, Decimals and Mixed Numbers

- ➢ Factoring Numbers, Greatest Common Factor, and Least Common

  Multiple

- ➢ Divisibility Rules

"A Man is like a fraction whose numerator is what he is and whose denominator is what he thinks of himself. The larger the denominator, the smaller the fraction." –Tolstoy

Name: ....

Simplifying Fractions

✓ Regularly divide both the top and bottom of the fraction by $2, 3, 5, 7, \ldots$ etc.
✓ Continue until you can't go any further.

EXAMPLE:

Simplify $\dfrac{12}{20}$.

To simplify $\dfrac{12}{20}$, you have to find a number that both 12 and 20 are divisible by. Both are divisible by 4. Then: $\dfrac{12}{20} = \dfrac{12 \div 4}{20 \div 4} = \dfrac{3}{5}$

PRACTICES:

Simplify the fractions.

1) $\dfrac{44}{64}$

2) $\dfrac{12}{26}$

3) $\dfrac{15}{25}$

4) $\dfrac{30}{45}$

5) $\dfrac{18}{27}$

6) $1\dfrac{62}{124}$

7) $4\dfrac{12}{66}$

8) $1\dfrac{55}{70}$

9) $\dfrac{54}{60}$

10) $7\dfrac{68}{136}$

Score: ....

	Answer Key	
1) $\frac{11}{16}$		2) $\frac{6}{13}$
3) $\frac{3}{5}$		4) $\frac{2}{3}$
5) $\frac{2}{3}$		6) $1\frac{1}{2}$
7) $4\frac{2}{11}$		8) $1\frac{11}{14}$
9) $\frac{9}{10}$		10) $7\frac{1}{2}$

Name: ....

Factoring Numbers

✓ To break the numbers into their prime factors is called factoring.

✓ First few prime numbers are $2, 3, 5, 7, 11, 13, 17, 19$

EXAMPLE:

List all positive factors of 12.

Write the upside-down division:

The second column is the answer.

Then: $12 = 2 \times 2 \times 3$ or $12 = 2^2 \times 3$

12	2
6	2
3	3
1	

PRACTICES:

List all positive factors of each number.	List the prime factorization for each number.
1) 90	2) 40
3) 49	4) 105
5) 50	6) 42
7) 34	8) 78
9) 96	10) 165

Score: ....

Answer Key	
1) 1, 2, 3, 5, 6, 9, 10, 15, 18, 30, 45, 90	2) $2 \times 2 \times 2 \times 5$
3) 1, 7, 49	4) $3 \times 5 \times 7$
5) 1, 2, 5, 10, 25, 50	6) $2 \times 3 \times 7$
7) 1, 2 , 17, 34	8) $2 \times 3 \times 13$
9) 1, 2, 3, 4, 6, 8, 12, 16, 24, 32, 48, 96	10) $3 \times 5 \times 11$

Name: ....

Greatest Common Factor (GCF)

✓ List the prime factors of each number.

✓ Then multiply common prime factors.

✓ If there are no common prime factors, then our GCF is 1.

EXAMPLE:

Find the GCF for **10** and **15**.

The factors of 10 are: $\{1, 2, 5, 10\}$

The factors of 15 are: $\{1, 3, 5, 15\}$

There is 5 in common,

Then the greatest common factor is: 5

PRACTICES:

Find the GCF for each number pair.

1) 12, 25	2) 72, 84
3) 24, 36	4) 30, 45
5) 9, 36	6) 63, 42
7) 27, 12	8) 125, 50
9) 54, 39	10) 36, 52

Score: ....

Answer Key	
1) 1	2) 12
3) 12	4) 15
5) 9	6) 21
7) 3	8) 25
9) 3	10) 4

Name: ....

Least Common Multiple (LCM)

✓ The smallest multiple that 2 or more numbers have in common is called least common multiple of that number. How to find LCM:

✓ First find the list of the prime factors of each number.

✓ Then multiply the common prime factors and uncommon prime factors of the numbers (each common prime factor is used only for once)

EXAMPLE:

Find the LCM for **18** and **12**.

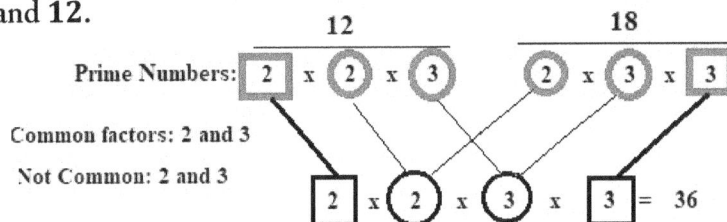

PRACTICES:

Find the LCM for each number pair.

1) 12, 9	2) 40, 20
3) 15, 30	4) 84, 60
5) 60, 40	6) 52, 78
7) 14, 28	8) 14, 7, 42
9) 24, 32	10) 72, 66, 24

Score: ....

Answer Key	
1) 36	2) 40
3) 30	4) 420
5) 120	6) 156
7) 28	8) 42
9) 96	10) 792

Name: ....

Divisibility Rules

If a number can be divided by other numbers, it is referred as divisibility. The number is divisible:

- ✓ by 2 if the number is found even.
- ✓ by 3 if the sum of the digits is found to be divisible by 3.
- ✓ by 9 if the sum of the digits is found to be divisible by 9.
- ✓ by 4, if the last 2 digits of a number are found to be divisible by 4.
- ✓ by 6, if it is found to be divisible by 2 and 3.
- ✓ by 8, if it is found to be divisible by 2 and 4.
- ✓ by 5 if the last digit is found 0 or 5.
- ✓ by 10 if the last digit is 0.

EXAMPLE:

What is the factor of 240?

2, because the number is even

3, because $(2 + 4 + 0 = 6, 6 \div 3 = 2)$

4, because $(40 \div 4 = 10)$

5, because the last digit is 0

8, because of 2 and 4

10, because the last digit is 0

Then 240 is divisible by 2, 3, 4, 5, 8, 10

PRACTICES:

Use the divisibility rules to find the factors of each number.

1) 12	2 3 4 5 6 7 8 9 10	2) 24	2 3 4 5 6 7 8 9 10
3) 36	2 3 4 5 6 7 8 9 10	4) 18	2 3 4 5 6 7 8 9 10
5) 30	2 3 4 5 6 7 8 9 10	6) 54	2 3 4 5 6 7 8 9 10
7) 90	2 3 4 5 6 7 8 9 10	8) 80	2 3 4 5 6 7 8 9 10
9) 72	2 3 4 5 6 7 8 9 10	10) 84	2 3 4 5 6 7 8 9 10

Score: ....

Answer Key		
1) 12	**2** **3** **4** 5 **6** 7 8 9 10	
2) 24	**2** **3** **4** 5 **6** 7 **8** 9 10	
3) 36	**2** **3** 4 5 **6** 7 8 **9** 10	
4) 18	**2** **3** 4 5 **6** 7 8 **9** 10	
5) 30	**2** **3** 4 **5** **6** 7 8 9 **10**	
6) 54	**2** **3** 4 5 **6** 7 8 **9** 10	
7) 90	**2** **3** 4 **5** **6** 7 8 **9** **10**	
8) 80	**2** 3 **4** **5** 6 7 **8** 9 **10**	
9) 72	**2** **3** **4** 5 **6** 7 **8** **9** 10	
10) 84	**2** **3** **4** 5 **6** **7** 8 9 10	

Name: ....

Adding and Subtracting Fractions

✓ Find equivalent fractions with the equivalent divisor before you can add or subtract fractions with totally different divisors.

✓ Adding and Subtracting with the equivalent divisors:

$$\frac{a}{b} + \frac{c}{b} = \frac{a+c}{b} \; , \frac{a}{b} - \frac{c}{b} = \frac{a-c}{b}$$

✓ Adding and Subtracting fractions with different divisors:

$$\frac{a}{b} + \frac{c}{d} = \frac{ad+cb}{bd} \; , \frac{a}{b} - \frac{c}{d} = \frac{ad-cb}{bd}$$

EXAMPLE:

Subtract fractions. $\frac{2}{3} - \frac{1}{2} = ?$

For "unlike" fractions, find equivalent fractions with the same divisors before you can add or subtract fractions with different divisors. Use this formula: $\frac{a}{b} - \frac{c}{d} = \frac{ad-cb}{bd}$

$$\frac{2}{3} - \frac{1}{2} = \frac{(2)(2) - (1)(3)}{3 \times 2} = \frac{4-3}{6} = \frac{1}{6}$$

PRACTICES:

Add fractions.	Subtract fractions.
1) $\frac{1}{4} + \frac{2}{3}$	2) $\frac{1}{2} - \frac{1}{5}$
3) $\frac{1}{3} + \frac{1}{2}$	4) $\frac{1}{7} - \frac{1}{9}$
5) $\frac{1}{4} + \frac{5}{7}$	6) $\frac{3}{5} - \frac{1}{15}$
7) $\frac{6}{7} + \frac{3}{21}$	8) $\frac{1}{3} - \frac{1}{4}$
9) $\frac{5}{13} + \frac{1}{2}$	10) $\frac{6}{5} - \frac{5}{6}$

Score: ....

	Answer Key	
1) $\frac{11}{12}$		2) $\frac{3}{10}$
3) $\frac{5}{6}$		4) $\frac{2}{63}$
5) $\frac{27}{28}$		6) $\frac{8}{15}$
7) 1		8) $\frac{1}{12}$
9) $\frac{23}{26}$		10) $\frac{11}{30}$

Name: ....

Multiplying and Dividing Fractions

- ✓ How to multiply fractions: multiply the top numbers and multiply the bottom numbers.
- ✓ How to change fractions: Keep, Change, Flip
- ✓ Keep the first fraction then change division sign into multiplication sign and flip the numerator and denominator of the second fraction. Then, solve it

EXAMPLE:

Multiplying fractions. $\frac{5}{6} \times \frac{3}{4} =$

Multiply the upper numbers and multiply the lower numbers.

$\frac{5}{6} \times \frac{3}{4} = \frac{5\times3}{6\times4} = \frac{15}{24}$, simplify: $\frac{15}{24} = \frac{15\div3}{24\div3} = \frac{5}{8}$

Dividing fractions. $\frac{1}{4} \div \frac{2}{3} =$

Keep the first fraction then change division sign into multiplication sign and flip the numerator and denominator of the second fraction. Then: $\frac{1}{4} \times \frac{3}{2} = \frac{1\times3}{4\times2} = \frac{3}{8}$

PRACTICES:

Multiplying fractions. Then simplify.	Dividing fractions.
1) $\frac{3}{5} \times \frac{5}{9}$	2) $\frac{4}{9} \div 4$
3) $\frac{5}{21} \times \frac{7}{10}$	4) $\frac{32}{25} \div \frac{8}{5}$
5) $\frac{3}{29} \times \frac{29}{3}$	6) $\frac{2}{7} \div \frac{8}{35}$
7) $\frac{8}{11} \times 11$	8) $\frac{12}{25} \div \frac{3}{5}$
9) $\frac{7}{9} \times \frac{12}{28}$	10) $7 \div \frac{2}{3}$

Score: ....

Answer Key	
1) $\frac{1}{3}$	2) $\frac{1}{9}$
3) $\frac{1}{6}$	4) $\frac{4}{5}$
5) 1	6) $1\frac{1}{4}$
7) 8	8) $\frac{4}{5}$
9) $\frac{1}{3}$	10) $10\frac{1}{2}$

Name: ....

Adding Mixed Numbers

Use these steps for both adding and subtracting mixed numbers.
- ✓ Add whole number of the mixed numbers.
- ✓ Add the fractions of every mixed number.
- ✓ Find the Least Common Divisor (LCD) if needed.
- ✓ Add whole numbers and fractions.
- ✓ Write your answer in simplest form.

EXAMPLE:

$1\frac{3}{4} + 2\frac{3}{8} = ?$

Rewriting our equation with parts separated, $1 + \frac{3}{4} + 2 + \frac{3}{8}$, Solving the whole number

parts $1 + 2 = 3$, Solving the fraction parts $\frac{3}{4} + \frac{3}{8}$, and rewrite to solve with the equivalent

fractions.

$\frac{6}{8} + \frac{3}{8} = \frac{9}{8} = 1\frac{1}{8}$, then combining the whole and fraction parts $3 + 1 + \frac{1}{8} = 4\frac{1}{8}$

PRACTICES:

Add.

1) $1\frac{1}{7} + 2\frac{1}{3}$	2) $1\frac{1}{2} + 3\frac{2}{3}$
3) $1\frac{2}{5} + 2\frac{1}{10}$	4) $7 + 2\frac{1}{2}$
5) $4\frac{1}{3} + 2\frac{2}{3}$	6) $2\frac{2}{3} + 1\frac{1}{4}$
7) $2\frac{3}{4} + 3\frac{1}{8}$	8) $9 + 1\frac{1}{9}$
9) $4\frac{5}{12} + 2\frac{3}{4}$	10) $3\frac{1}{7} + 2\frac{3}{14}$

Score: ....

Answer Key

1) $3\frac{10}{21}$	2) $5\frac{1}{6}$
3) $3\frac{1}{2}$	4) $9\frac{1}{2}$
5) 7	6) $3\frac{11}{12}$
7) $5\frac{7}{8}$	8) $10\frac{1}{9}$
9) $7\frac{1}{6}$	10) $5\frac{5}{14}$

Name: ....

Subtracting Mixed Numbers

Use these steps for both adding and subtracting mixed numbers.
- ✓ From whole number of the first mixed number, subtract the whole number of second mixed number.
- ✓ From first fraction subtract the second.
- ✓ Find the Least Common Divisor (LCD) if needed.
- ✓ Add the result of whole numbers and fractions.
- ✓ Write your answer in simplest terms.

EXAMPLE:

$5\frac{2}{3} - 2\frac{1}{4} = ?$

Rewriting our equation with parts separated, $5 + \frac{2}{3} - 2 - \frac{1}{4}$

Solving the whole number parts $5 - 2 = 3$, Solving the fraction parts, $\frac{2}{3} - \frac{1}{4} = \frac{8-3}{12} = \frac{5}{12}$

Joining the whole and fraction parts, $3 + \frac{5}{12} = 3\frac{5}{12}$

PRACTICES:

Subtract.

1) $5\frac{2}{7} - 2\frac{1}{14}$	2) $4\frac{2}{5} - \frac{2}{3}$
3) $3\frac{3}{7} - 1\frac{1}{14}$	4) $7\frac{2}{5} - 5\frac{1}{3}$
5) $4\frac{1}{2} - 2\frac{4}{8}$	6) $11\frac{5}{12} - 8\frac{3}{4}$
7) $7\frac{5}{12} - 5\frac{7}{12}$	8) $5\frac{2}{9} - 2\frac{1}{18}$
9) $3\frac{2}{5} - 2\frac{1}{5}$	10) $3\frac{4}{9} - 1\frac{2}{9}$

Score: ....

Answer Key	
1) $3\frac{3}{14}$	2) $3\frac{11}{15}$
3) $2\frac{5}{14}$	4) $2\frac{1}{15}$
5) 2	6) $2\frac{2}{3}$
7) $1\frac{5}{6}$	8) $3\frac{1}{6}$
9) $1\frac{1}{5}$	10) $2\frac{2}{9}$

Name: ....

Multiplying Mixed Numbers

✓ Convert the mixed numbers into improper fractions. (Improper fraction is a fraction in which the numerator is greater than denominator)

✓ Multiply fractions and write in simplest form if needed.

$$a\frac{c}{b} = a + \frac{c}{b} = \frac{ab + c}{b}$$

EXAMPLE:

Multiply mixed numbers. $4\frac{3}{5} \times 2\frac{1}{3} = ?$

Changing mixed numbers to fractions, $\frac{23}{5} \times \frac{7}{3}$, Applying the fractions formula for multiplication, $\frac{23 \times 7}{5 \times 3} = \frac{161}{15} = 10\frac{11}{15}$

PRACTICES:

Find each product.

1) $2\frac{1}{3} \times \frac{1}{2}$

2) $1\frac{2}{5} \times \frac{2}{3}$

3) $2\frac{4}{3} \times 2\frac{2}{6}$

4) $2\frac{1}{2} \times 1\frac{2}{4}$

5) $3\frac{1}{2} \times 1\frac{2}{3}$

6) $1\frac{1}{7} \times 1\frac{3}{4}$

7) $1\frac{1}{4} \times 2\frac{6}{5}$

8) $3\frac{1}{2} \times 4\frac{2}{5}$

9) $1\frac{2}{5} \times 2\frac{1}{3}$

10) $5\frac{7}{12} \times 2\frac{4}{9}$

Score: ....

Answer Key	
1) $1\frac{1}{6}$	2) $\frac{14}{15}$
3) $7\frac{7}{9}$	4) $3\frac{3}{4}$
5) $5\frac{5}{6}$	6) 2
7) 4	8) $15\frac{2}{5}$
9) $3\frac{4}{15}$	10) $13\frac{35}{54}$

Name: ....

Dividing Mixed Numbers

✓ Change the mixed numbers into improper fractions.

✓ Divide fractions and write in simplest form if needed.

$$a\frac{c}{b} = a + \frac{c}{b} = \frac{ab + c}{b}$$

EXAMPLE:

Find the quotient. $2\frac{1}{2} \div 1\frac{1}{5} = ?$

Changing mixed numbers to fractions, $\frac{5}{2} \div \frac{6}{5}$, Applying the fractions formula for

multiplication, $\frac{5}{2} \times \frac{5}{6} = \frac{5 \times 5}{2 \times 6} = \frac{25}{12} = 2\frac{1}{12}$

PRACTICES:

Find each quotient.

1) $2\frac{3}{5} \div 1\frac{3}{8}$	2) $\frac{3}{2} \div 2\frac{3}{4}$
3) $1\frac{4}{7} \div 2\frac{2}{3}$	4) $1\frac{2}{3} \div 2\frac{1}{3}$
5) $0 \div 4\frac{2}{5}$	6) $2\frac{2}{5} \div 1\frac{1}{2}$
7) $1\frac{2}{3} \div 2\frac{1}{5}$	8) $3\frac{2}{7} \div 4\frac{3}{5}$
9) $1\frac{1}{4} \div 2\frac{4}{5}$	10) $2 \div 3\frac{1}{3}$

Score: ....

Answer Key

1) $1\frac{49}{55}$	2) $\frac{6}{11}$
3) $\frac{33}{56}$	4) $\frac{5}{7}$
5) 0	6) $1\frac{3}{5}$
7) $\frac{25}{33}$	8) $\frac{5}{7}$
9) $\frac{25}{56}$	10) $\frac{3}{5}$

Name: ....

Comparing Decimals

Decimal is a fraction written in a unique form. For example, instead of writing $\frac{1}{2}$ you can write as **0.5**.

For comparison of decimals:

✓ Compare every digit of two decimals in the same place value.

✓ Start from left. Compare ones, tens, hundreds, tenths, hundredths, etc.

✓ To compare numbers, use these symbols:

- Equal to =, Less than <,   Greater than >

- Greater than or equal ≥,   Less than or equal ≤

EXAMPLE:

Compare **0.20** and **0.02**.

0.20 *is greater than* 0.02, because the tenth place of 0.20 is 2, but the tenth place of 0.02 is zero. Then: 0.20 > 0.02

PRACTICES:

Write the correct comparison symbol (>, < or =).

1) 0.025 _____ 0.25	2) 0.9 _____ 0.888
3) 4.510 _____ 4.150	4) 10.01 _____ 10.10
5) 0.987 _____ 0.991	6) 18.004 _____ 18.040
7) 0.020 _____ 0.20	8) 0.071___ _0.700
9) 0.08____0.009	10) 0.690 _____ 0.609

Score: ....

Answer Key	
1) <	2) >
3) >	4) <
5) <	6) <
7) <	8) <
9) >	10) >

Name: ....

Rounding Decimals

- ✓ To round a decimal, you must find the place value you'll round to.
- ✓ Then, find the digit to the right of the place value you're rounding to.
  - If it is 5 or greater, add 1 to the place value you're rounding to and remove all digits on its right side.
  - If the digit to the right of the place value is smaller than 5, keep the place value and remove all digits on the right.

EXAMPLE:

Round 2.1837 to the thousandth-place value.

First have a look at the next place value to the right, (tens thousandths). It's 7 and it's found to be greater than 5. So, add 1 to the digit in the thousandth place.

Thousandth place is 3. → 3 + 1 = 4, then, the answer is 2.184

PRACTICES:

Round each decimal number to the nearest place indicated.

1) 6.0<u>8</u>	2) 12.<u>2</u>67
3) 9.<u>3</u>01	4) 10.0<u>7</u>1
5) 5<u>5</u>.89	6) 5<u>9</u>.15
7) 3<u>2</u>9.018	8) 92.<u>4</u>10
9) 1.4<u>9</u>9	10) 2<u>5</u>.621

Score: ....

Answer Key	
1) 6.1	2) 12.3
3) 9.3	4) 10.07
5) 56	6) 59
7) 330	8) 92.4
9) 1.5	10) 26

Name: ....

Adding and Subtracting Decimals

- ✓ Arrange the numbers in line.
- ✓ Add zeros to have same number of digits for both the numbers.
- ✓ Add or subtract by using column subtraction or addition.

EXAMPLE:

Add. $2.5 + 1.24 =$

First line up the numbers: $\frac{2.5}{+1.24}$ → Add zeros to have same number of digits for both

numbers. $\frac{2.50}{+1.24}$, Start with the hundredths place. $0 + 4 = 4$, $\frac{2.50}{+1.24}{4}$, Continue with tenths

place. $5 + 2 = 7$, $\frac{2.50}{+1.24}{.74}$. Add the ones place. $2 + 1 = 3$, $\frac{2.50}{+1.24}{3.74}$

Subtract decimals. $4.67 - 2.15 = \frac{4.67}{-2.15}$

Start with the hundredths place. $7 - 5 = 2$, $\frac{4.67}{-2.15}{2}$, continue with tenths place. $6 - 1 = 5$

$\frac{4.67}{-2.15}{.52}$, subtract the ones place. $4 - 2 = 2$, $\frac{4.67}{-2.15}{2.52}$.

PRACTICES:

Add and subtract decimals.

1)	$\frac{87.15}{-32.35}$	2)	$\frac{90.43}{+44.09}$
3)	$\frac{58.56}{+12.10}$	4)	$\frac{65.23}{-56.48}$
5)	$\frac{98.125}{+58.54}$	6)	$\frac{162.05}{-83.65}$

Solve.

7) ___ $+ 5.0 = 9.08$	8) $7.06 + $ ___ $= 24.6$
9) $21.9 - $ ___ $= 6.9$	10) $32.12 - $ ___ $= 12.07$

Score: ....

Answer Key	
1) 54.8	2) 134.52
3) 70.66	4) 8.75
5) 156.665	6) 78.4
7) 4.08	8) 17.54
9) 15	10) 20.05

Name: ....

Multiplying Decimals

✓ Arrange and multiply the numbers as you do with whole numbers.
✓ Then count the total number of decimal places in every factor.
✓ Place the decimal point in the product.

EXAMPLE:

Find the product. $0.50 \times 0.20 =$

Arrange and multiply the numbers as you do with whole numbers. Line up the numbers:

$\begin{array}{r} 50 \\ \times\,20 \\ \hline \end{array}$, Start with the ones place $\to 0 \times 50 = 0$, $\begin{array}{r} 50 \\ \times 20 \\ \hline 0 \end{array}$, Continue with other digits $\to$

$2 \times 50 = 100$, $\begin{array}{r} 50 \\ \times 20 \\ \hline 1,000 \end{array}$, Count the total number of decimal places in both of the factors

(4). Then Place the decimal point in the product.

Then: $\begin{array}{r} 0.50 \\ \times\,0.20 \\ \hline 0.1000 \end{array} \to 0.50 \times 0.20 = 0.1$

PRACTICES:

Find each product.

1) $\begin{array}{r} 1.5 \\ \times\,0.16 \\ \hline \end{array}$	2) $\begin{array}{r} 5.3 \\ \times\,1.9 \\ \hline \end{array}$
3) $\begin{array}{r} 0.06 \\ \times\,2.5 \\ \hline \end{array}$	4) $\begin{array}{r} 3.19 \\ \times\,21.5 \\ \hline \end{array}$
5) $\begin{array}{r} 9.3 \\ \times\,11.5 \\ \hline \end{array}$	6) $\begin{array}{r} 3.01 \\ \times\,2.1 \\ \hline \end{array}$
7) $\begin{array}{r} 5.0 \\ \times\,1.4 \\ \hline \end{array}$	8) $\begin{array}{r} 23.8 \\ \times\,10 \\ \hline \end{array}$
9) $\begin{array}{r} 21.5 \\ \times\,0.001 \\ \hline \end{array}$	10) $\begin{array}{r} 8.21 \\ \times\,3.1 \\ \hline \end{array}$

Score: ....

Answer Key	
1) 0.24	2) 10.07
3) 0.15	4) 68.585
5) 106.95	6) 6.321
7) 7	8) 238
9) 0.0215	10) 25.451

Name: ....

Dividing Decimals

- ✓ If the divisor is not a whole number, transfer decimal point to right to make it a whole number. Do the same step for dividend.
- ✓ Divide same to whole numbers.

EXAMPLE:

Find the quotient. $1.20 \div 0.2 =$

The divisor is not a whole number. Multiply it by 10 to get 2. Do the same step for the dividend to get 12. Now, divide: $12 \div 2 = 6$. The answer is 6.

PRACTICES:

Find each quotient.

1) $25.7 \div 0.5$	2) $67.2 \div 4$
3) $61.75 \div 1.9$	4) $18.0 \div 1.2$
5) $12.4 \div 10$	6) $2.2 \div 100$
7) $1.88 \div 100$	8) $55.1 \div 100$
9) $0.1 \div 100$	10) $0.25 \div 10$

Score: ....

Answer Key	
1) 51.4	2) 16.8
3) 32.5	4) 15
5) 1.24	6) 0.022
7) 0.0188	8) 0.551
9) 0.001	10) 0.025

Name: ....

Converting Between Fractions, Decimals and Mixed

How to convert fraction into Decimal:

✓ Divide the numerator by denominator.

How to convert decimal into Fraction:

✓ Write decimal over 1.

✓ Multiply both numerator value and denominator value by 10 for each digit on the right side of the decimal point.

✓ Make it to simplest form.

EXAMPLE:

What is long division of $\frac{5}{8}$ =?

In that case we put extra zeros and did $\frac{5.000}{8}$ to get 0.625

PRACTICES:

Convert fractions to decimals.	Convert decimal into fraction or mixed numbers
1) $\frac{4}{10}$	2) 3.6
3) $\frac{3}{8}$	4) 0.07
5) $\frac{4}{12}$	6) 0.15
7) $\frac{5}{16}$	8) 2.7
9) $\frac{60}{100}$	10) 2.5

Score: ....

Answer Key	
1) 0.4	2) $3\frac{3}{5}$
3) 0.375	4) $\frac{7}{100}$
5) 0.333	6) $\frac{3}{20}$
7) 0.3125	8) $2\frac{7}{10}$
9) 0.6	10) $2\frac{1}{2}$

Chapter 3 : Proportion, Ratio, Percent

Topics that you'll learn in this chapter:

- ➢ Writing and Simplifying Ratios

- ➢ Create a Proportion

- ➢ Similar Figures

- ➢ Simple Interest

- ➢ Ratio and Rates Word Problems

- ➢ Percentage Calculations

- ➢ Converting Between Percent, Fractions, and Decimals

- ➢ Percent Problems

- ➢ Markup, Discount, and Tax

"Do not worry about your difficulties in mathematics. I can assure you mine are still greater." – *Albert Einstein*

Name: ....

Writing Ratios

✓ A ratio is a comparison of two numbers, and it can be written as a division.

EXAMPLE:

$3 : 5 = ?$

Both numbers 3 and 5 are divisible by 8 , $\Rightarrow 3 \div 8 = \frac{3}{8}, 5 \div 8 = \frac{5}{8},$

Then: $3 : 5 = \frac{3}{8}$ and $\frac{5}{8}$.

PRACTICES:

Express each ratio as a rate and unite rate.	Express each ratio as a fraction in the simplest form
1) 80 dollars for 4 chairs.	2) 13 cups to 39 cups.
3) 125miles on 25 gallons of gas.	4) 17 cakes out of 51 cakes
5) 147 miles on 7 hours.	6) 35 red desks out of 125 desks
7) 12 inches of snow in 24 hours.	8) 8 story books out of 32 books
9) 14 dimes to 112 dimes.	10) 12 gallons to 20 gallons

Score: ....

Answer Key	
1) $\frac{80 \text{ dollars}}{4 \text{ books}}$, 20.00 dollars per chair	2) $\frac{1}{3}$
3) $\frac{125 \text{ miles}}{25 \text{ gallons}}$, 5 miles per gallon	4) $\frac{1}{3}$
5) $\frac{147 \text{ miles}}{7 \text{ hours}}$, 21 miles per hour	6) $\frac{7}{25}$
7) $\frac{12" \text{ of snow}}{24 \text{ hours}}$, 0.5 inches of snow per hour	8) $\frac{1}{4}$
9) $\frac{14 \text{ dimes}}{112 \text{ dimes}}$, $\frac{1}{8}$ per dime	10) $\frac{3}{5}$

Name: ....

Simplifying Ratios

- ✓ Ratios are used to compare two numbers.
- ✓ Ratios can be written as a fraction, using colon or the word "to".
- ✓ You can calculate identical ratios by multiplying or dividing both sides of the ratio by the same number.

EXAMPLE:

Simplify. $8:4 =$

Both numbers 8 and 4 are divisible by 4 , $\Rightarrow 8 \div 4 = 2, 4 \div 4 = 1,$

Then: $8:4 = 2:1$

PRACTICES:

Reduce each ratio.

1) 49: 14	2) 22: 55
3) 35: 25	4) 18: 99
5) 16: 36	6) 64: 72
7) 4: 60	8) 70: 40
9) 8: 64	10) 16: 24

Score: ....

Answer Key	
1) 7: 2	2) 2: 5
3) 7: 5	4) 2: 11
5) 4: 9	6) 8: 9
7) 1: 15	8) 7: 4
9) 1: 8	10) 2: 3

Name: ....

Create a Proportion

- ✓ A proportion carries two equal fractions! A proportion means equality of two fractions.
- ✓ If you want to create a proportion, simply find (or create) two equal fractions.

EXAMPLE:

Explain if these ratios form a proportion. $\frac{3}{5}$ and $\frac{24}{45}$

Use cross multiplication: $\frac{3}{5} = \frac{24}{45} \rightarrow 3 \times 45 = 5 \times 24 \rightarrow 135 = 120$, which is not correct.

Thus, this pair of ratios doesn't form a proportion.

PRACTICES:

Create proportion from the given set of numbers.

1) 3, 2, 9, 6	2) 4, 18, 12, 6
3) 5, 11, 25, 55	4) 24, 7, 21, 8
5) 49, 7, 12, 84	6) 15, 12, 30, 24
7) 20, 10, 200, 1	8) 9, 27, 81, 3
9) 4, 2, 16, 32	10) 9, 6, 27, 18

Score: ....

Answer Key	
1) 2: 6 = 3: 9	2) 4: 12 =6: 18
3) 5: 25 = 11: 55	4) 8: 24 =7: 21
5) 7: 49 = 12: 84	6) 12: 24 =15: 30
7) 1: 10 = 20: 200	8) 3: 27 =9: 81
9) 2: 16=4: 32	10) 6: 18 = 9: 27

Name: ....

Similar Figures

✓ Two or more figures are equivalent if their corresponding angles are equal, and the corresponding sides are in proportion.

EXAMPLE:

4–5–6 triangle is like an 8–10–12 triangle.

PRACTICES:

Each pair of figures is similar. Find the missing side.

1)

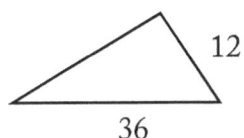 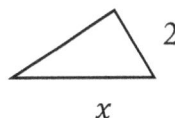

2)

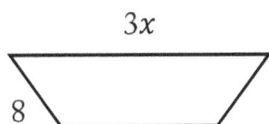 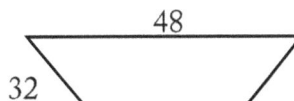

3)

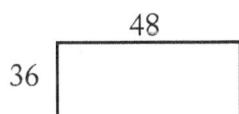 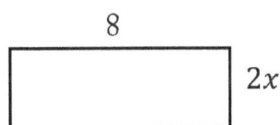

4)

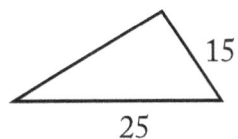 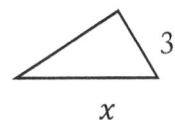

5)

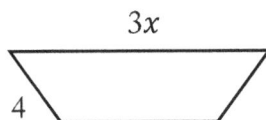 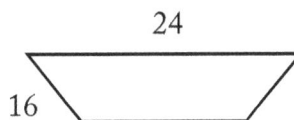

6)

```
        2x                              36
  ┌──────────┐                   ┌──────────┐
 4 └────────────┐          24   └────────────┐
```

7)

```
        63                              7
  ┌────────────┐              ┌────────────┐
54│            │              │            │6x
  └────────────┘              └────────────┘
```

8)

```
          27                         3
       /\                          /\
      /  \ 72                     /  \ x
```

9)

```
        4                              2x
   ┌────────┐                    ┌──────────┐
 6 └──────────┐              60 └────────────┘
```

10)

```
        45                              5
  ┌────────────┐              ┌──────────┐
  │            │x             │          │2
  └────────────┘              └──────────┘
```

Score: ....

Answer Key	
1) 6	2) 4
3) 3	4) 5
5) 2	6) 3
7) 1	8) 8
9) 20	10) 18

Name: ....

Ratio and Rates Word Problems

✓ To solve a rate word problem or a ratio, create a proportion and then use cross multiplication method.

EXAMPLE:

A tree **32 feet** tall has a shadow **12 feet** long. Jack is **6 feet** tall. How long is Jack's shadow?

To solve for the missing number, write in a proportion.

$\frac{32}{12} = \frac{6}{x} \rightarrow 32x = 6 \times 12 = 72$

$32x = 72 \rightarrow x = \dfrac{72}{32} = 2.25$

PRACTICES:

Solve.

1) In a party, 8 soft drinks are required for every 35guests. If there are 560 guests, how many soft drinks is required?

2) You can buy 6 cans of green beans at a supermarket for $3.50. How much does it cost to buy 42 cans of green beans?

3) The price of 5 bananas at the first Market is $1.05. The price of 7of the same bananas at second Market is $1.07. Which place is the better buy?

4) In Peter's class, 21 of the students are tall and 9 are short. In Elise's class 56 students are tall and 24 students are short. Which class has a higher ratio of tall to short students?

5) The bakers at a Bakery can make 110 bagels in 4 hours. How many bagels can they bake in 6 hours? What is that rate per hour?

6) A certain sweet recipe calls for 3 kg of sugar for every 6 kg of flour. If 63 kg of this sweet must be prepared, how much sugar is required?

7) In a mixture of 45 liters, the ratio of sugar solution to salt solution is 1:2. What is the amount of sugar solution to be added if the ratio must be 2:1?

8) In a bag of red and green sweets, the ratio of red sweets to green sweets is 3:4. If the bag contains 120 green sweets, how many red sweets are there?

9) If the ratio of chocolates to ice-cream cones in a box is 5:8 and the number of chocolates is 30, find the number of ice-cream cones.

10) In a group, the ratio of doctors to lawyers is 5:4. If the total number of people in the group is 72, what is the number of lawyers in the group?

Score: ....

Answer Key

1) 128	2) $24.5
3) The price at the second Market is a better buy.	4) The ratio for both classes equal 7 to 3.
5) 165, the rate is 27.5 per hour.	6) 21 kg (3+6=9, $\frac{63}{9} = 7$. Therefore, 3:6=21:42)
7) 45	8) 90
9) 48	10) 32

Name: ....

Percentage Calculations

- ✓ Percent is called the ratio of a number and 100. It always possesses the same denominator, 100. The symbol used for percent is %.
- ✓ Percent is another method to write decimals or fractions. For example:

  $$40\% = 0.40 = \frac{40}{100} = \frac{2}{5}$$

- ✓ Use the given formula to find part, whole, or percent:

  $$\text{part} = \frac{\text{percent}}{100} \times \text{whole}$$

EXAMPLE:

What is **10% of 45?**

Use this formula: $\text{part} = \frac{\text{percent}}{100} \times \text{whole}$

$$\text{part} = \frac{10}{100} \times 45 \rightarrow \text{part} = \frac{1}{10} \times 45 \rightarrow \text{part} = \frac{45}{10} \rightarrow \text{part} = 4.5$$

PRACTICES:

Calculate the percentages.

1) 75% of 45	2) 50% of 66
3) 90% of 58	4) 25% of 88
5) 5% of 100	6) 80% of 60

Solve.

7) What percentage of 60 is 6	8) 6.76 is what percentage of 52?
9) 17 is what percentage of 85?	10) Find what percentage of 96 is 24.

Score: ....

Answer Key	
1) 33.75	2) 33
3) 52.2	4) 22
5) 5	6) 48
7) 10%	8) 13%
9) 20%	10) 25%

Name: ....

Percent Problems

✓ In each percent question, we are finding the base, or part or the percent.
✓ Use the following equations to find each missing portion.
- Base = Part ÷ Percent
- Part = Percent × Base
- Percent = Part ÷ Base

EXAMPLE:

20 is 5% of what number?

Use the formula: $Base = Part \div Percent \rightarrow Base = 20 \div 0.05 = 400$

20 is 5% of 400

PRACTICES:

Solve each problem.

1) 52% of what number is 13?	2) What is 15% of 9 inches?
3) What percent of 185.6 is 23.2?	4) 24 is 72% of what?
5) 35 is what percent of 70?	6) 10 is 200% of what?
7) 14 is what percent of 70?	8) 26% of 100 is what number?

9) Mia requires 50% to pass. If she gets 250 marks and falls short by 90 marks, what were the maximum marks she could have got?

10) Jack scored 14 out of 70 marks in mathematics, 9 out of 10 marks in history and 56 out of 100 marks in science. In which subject his percentage of marks is the best?

Score: ....

Answer Key	
1) 25	2) 1.35
3) 12.5	4) 33.33
5) 50%	6) 5
7) 20%	8) 26
9) 680	10) History

Name: ....

Markup, Discount, and Tax

- ✓ Markup = selling price – cost
- ✓ Markup rate = markup is divided by the cost
- ✓ Discount = Multiply the rate of discount by regular price.
- ✓ Tax: To find tax, multiply the taxable amount (income, property value, etc.) to the tax rate.
- ✓ To find tip, multiply selling price to the rate.

EXAMPLE:

With an **10%** discount, Ella was able to save **$20** on a dress. What was the original price of the dress?

$$10\% \ of \ x = \ 20, \frac{10}{100} \times x = \ 20, x = \frac{100 \times 20}{10} = 200$$

PRACTICES:

Find the selling price of each item.

1) Cost of a chair: $20, markup: 30%, discount: 10%, tax: 10%

2) Cost of computer: $1,600.00, markup: 65%

3) Cost of a pen: $3.20, markup: 50%, discount: 15%, tax: 5%

4) Cost of a puppy: $1,800, markup: 40%, discount: 10%

5) Cost of a book: $50, markup: 40%, discount: 20%, tax: 5%

6) Original price of a tablet: $400, discount: 20% Tax: 5%,

7) Original price of a book: $50, markup:20% Discount: 20%, Tax: 2.5%,

8) Original price of a cellphone: $500, markup:14% Discount: 25%, Tax: 1.6%,

9) Original price of a sofa: $800, markup:10% Discount: 15%, Tax: 1.5%,

10) Original price of a car: $40,000, markup:12% Discount: 25%, Tax: 6.5%,

Answer Key

1) $25.74	2) $2,640
3) $4.284	4) $2,268
5) $58.8	6) $336
7) $49.2	8) $434.34
9) $759.22	10) $35,784

Name: ....

Simple Interest

✓ Simple Interest: The charge for borrowing money or the return for lending it.

To solve a simple interest problem, use this formula:

✓ Interest = principal × rate × time $\Rightarrow I = p \times r \times t$

EXAMPLE:

Find simple interest for $5,200 at 4% for 3 years.

Use Interest formula: $I = prt$

$P = \$5,200$, r = 4% = $\frac{4}{100}$ = 0.04 and $t = 3$

Then: $I = 5,200 \times 0.04 \times 3 = \$624$

PRACTICES:

Use simple interest to find the ending balance.

1) $1,200 at 15% for 3 years.	2) $320,000 at 2.85% for 7 years.
3) $1,500 at 2.25% for 12 years.	4) $12,500 at 6.2% for 4 years.
5) $31,000 at 1.5% for 10 months.	6) $18,000 at 5.2% for 5 years.

7) Emily puts $6,000 into an investment yielding 3.25% annual simple interest; she left the money in for 3 years. How much interest does Sara get at the end of those 3 years?

8) A new car, valued at $42,000, depreciates at 7.5% per year from original price. Find the value of the car 6 years after purchase.

9) $880 interest is earned on a principal of $2,200 at a simple interest rate of 4% interest per year. For how many years was the principal invested?

10) A bank is offering 3.2% simple interest on a savings account. If you deposit $15,000, how much interest will you earn in six years?

Score: ....

Answer Key

1) $1,740	2) $383,840.00
3) $1,905.00	4) $15,600
5) $31,387.50	6) $22,680
7) $585.00	8) $23,100
9) 10 years	10) $2,880

Name: ....

Converting Between Percent, Fractions, and Decimals

✓ To a percent: We move the decimal point 2 places to the right and add the percentage (%) symbol.

✓ Divide by 100 to change a number from percent to decimal.

EXAMPLE:

30% $= 0.30$

0.24 $= 24\%$

PRACTICES:

Converting fractions to decimals.	Write each decimal as a percent.
1) $\frac{23}{10}$	2) 0.002
3) $\frac{2}{20}$	4) 0.08
5) $\frac{7}{100}$	6) 0.2
7) $\frac{20}{50}$	8) 3.25
9) $\frac{3}{60}$	10) 1.01

Score: ....

Answer Key	
1) 2.3	2) 0.2%
3) 0.1	4) 8%
5) 0.07	6) 20%
7) 0.4	8) 325%
9) 0.05	10) 101%

Chapter 4 : Exponents and Radicals

Topics that you'll learn in this chapter:

> ➢ Multiplication Property of Exponents

> ➢ Division Property of Exponents

> ➢ Powers of Products and Quotients

> ➢ Zero, Negative Exponents and Bases

"Mathematics is no more computation than typing is literature." – *John Allen Paulos*

Name: ....

Multiplication Property of Exponents

- ✓ Exponents are shorthand for recurrent multiplication of the identical number by itself. For example, instead of writing 2×2, we can write 2^2. For $3 \times 3 \times 3 \times 3$, we can write 3^4

- ✓ In algebra, a variable is a letter used as a replacement for a number. The most common letters are: $x, y, z, a, b, c, m,$ and n.

- ✓ Exponent's rules: $(x^a)^b = x^{a \times b}$, $(xy)^a = x^a \times y^a$,

    $x^a \times x^b = x^{a+b}$, $x^a \times y^a = (xy)^a$,

EXAMPLE:

Multiply. $-2x^5 \times 7x^3 =$

   Use Exponent's rules: $x^a \times x^b = x^{a+b} \rightarrow x^5 \times x^3 = x^{5+3} = x^8$

   Then: $-2x^5 \times 7x^3 = -14x^8$

PRACTICES:

Simplify.

1) $4^3 \times 4^2$	2) $2 \times 2^2 \times 2^3$
3) $2^4 \times 2$	4) $8x^2 \times x$
5) $15x^7 \times x$	6) $3x \times x^3$
7) $2x^5 \times 5x^4$	8) $5x^2 \times 3x^2y^2$
9) $6y^5 \times 8xy^2$	10) $5xy^3 \times 4x^3y^2$

Score: ....

Answer Key	
1) 4^5	2) 2^6
3) 2^5	4) $8x^3$
5) $15x^8$	6) $3x^4$
7) $10x^9$	8) $15x^4y^2$
9) $48xy^7$	10) $20x^4y^5$

Name: ....

Division Property of Exponents

✓ For division of exponents, we can use these formulas: $\left(\frac{a}{b}\right)^c = \frac{a^c}{b^c}$, $b \neq 0$

$$\frac{x^a}{x^b} = x^{a-b}, x \neq 0 \qquad \frac{x^a}{y^a} = \left(\frac{x}{y}\right)^a, y \neq 0$$

$$\frac{x^a}{x^b} = \frac{1}{x^{b-a}}, x \neq 0, \qquad \frac{1}{x^b} = x^{-b}$$

EXAMPLE:

Simplify. $\frac{4x^3y}{36x^2y^3} =$

First you cancel the common factor: $4 \to \frac{4x^3y}{36x^2y^3} = \frac{x^3y}{9x^2y^3}$

Use Exponent's rules: $\frac{x^a}{x^b} = x^{a-b} \to \frac{x^3}{x^2} = x^{3-2}$

Then: $\frac{4x^3y}{36x^2y^3} = \frac{xy}{9y^3} \to$ now cancel the common factor: $y \to \frac{xy}{9y^3} = \frac{x}{9y^2}$

PRACTICES:

Simplify.

1) $\frac{4^3}{4}$

2) $\frac{51}{51^{14}}$

3) $\frac{5^2}{5^3}$

4) $\frac{3^4}{15^4}$

5) $\frac{x}{x^7}$

6) $\frac{42x^2}{6x^2}$

7) $\frac{3x^{-3}}{12x^{-1}}$

8) $\frac{81x^5}{9x^3}$

9) $\frac{3x^4}{4x^5}$

10) $\frac{21x}{3x^2}$

Score: ....

Answer Key	
1) 4^2	2) $\dfrac{1}{51^{13}}$
3) $\dfrac{1}{5}$	4) $\left(\dfrac{1}{5}\right)^4 = \dfrac{1}{5^4}$
5) $\dfrac{1}{x^6}$	6) 7
7) $\dfrac{1}{4x^2}$	8) $9x^2$
9) $\dfrac{3}{4x}$	10) $\dfrac{7}{x}$

Name: ....

Powers of Products and Quotients

✓ For any number except zero, a and b and any integer x, $(ab)^x = a^x \times b^x$.

EXAMPLE:

Simplify. $(3x^5y^4)^2 =$

Use Exponent's rules: $(x^a)^b = x^{a \times b}$

$$(3x^5y^4)^2 = (3)^2(x^5)^2(y^4)^2 = 9x^{5 \times 2}y^{4 \times 2} = 9x^{10}y^8$$

PRACTICES:

Simplify.

1) $(5x^3)^2$	2) $(xy)^2$
3) $(ax^2)^3$	4) $(2x^3yz)^2$
5) $(4x^2y^3)^2$	6) $(5x^2y^3)^2$
7) $(2xy^2)^3$	8) $(2x^3y)^4$
9) $(7x^4y^8)^2$	10) $(10x)^3$

Score: ....

Answer Key	
1) $25x^6$	2) x^2y^2
3) a^3x^6	4) $4x^6y^2z^2$
5) $16x^4y^6$	6) $25x^4y^6$
7) $8x^3y^6$	8) $8x^{12}y^4$
9) $49x^8y^{16}$	10) $1,000x^3$

Name: ....

Zero and Negative Exponents

✓ A negative exponent just means that the base is on the wrong side of the fraction line, so you need to flip the base to the other side. For example, "x^{-2}" (pronounced as "ecks to the minus two") just means "x^2" but below, as in $\frac{1}{x^2}$

EXAMPLE:

Evaluate. $\left(\frac{4}{9}\right)^{-2} =$

Use Exponent's rules: $\frac{1}{x^b} = x^{-b} \rightarrow \left(\frac{4}{9}\right)^{-2} = \frac{1}{\left(\frac{4}{9}\right)^2} = \frac{1}{\frac{4^2}{9^2}}$

Now use fraction rule: $\frac{1}{\frac{b}{c}} = \frac{c}{b} \rightarrow \frac{1}{\frac{4^2}{9^2}} = \frac{9^2}{4^2} = \frac{81}{16}$

PRACTICES:

Evaluate the following expressions.

1) 4^{-2}	2) 5^{-2}
3) 6^{-2}	4) 3^{-4}
5) 10^{-1}	6) 33^{-1}
7) 6^{-1}	8) 3^{-2}
9) 9^{-2}	10) 4^{-1}

Score: ....

	Answer Key	
1) $\frac{1}{16}$		2) $\frac{1}{25}$
3) $\frac{1}{36}$		4) $\frac{1}{81}$
5) $\frac{1}{10}$		6) $\frac{1}{33}$
7) $\frac{1}{6}$		8) $\frac{1}{9}$
9) $\frac{1}{81}$		10) $\frac{1}{4}$

Name: ....

Negative Exponents and Negative Bases

✓ First make the power positive. A negative exponent can be written as reciprocal of that number with a positive exponent.

✓ The parenthesis is significant!

✓ 5^{-3} is not the same as $(-5)^{-3}$

$$(-5)^{-3} = -\frac{1}{5^3} \text{ and } (5)^{-3} = +\frac{1}{5^3}$$

EXAMPLE:

Simplify. $(-\frac{5x}{3yz})^{-3} =$

Use Exponent's rules: $\frac{1}{x^b} = x^{-b} \rightarrow (-\frac{5x}{3yz})^{-3} = \frac{1}{(-\frac{5x}{3yz})^3} = \frac{1}{-\frac{5^3 x^3}{3^3 y^3 z^3}}$

Now use fraction rule: $\frac{1}{\frac{b}{c}} = \frac{c}{b} \rightarrow \frac{1}{\frac{5^3 x^3}{3^3 y^3 z^3}} = -\frac{3^3 y^3 z^3}{5^3 x^3} = -\frac{27 y^3 z^3}{125 x^3}$

PRACTICES:

Simplify.

1) 7^{-1}	2) $-2x^{-2}$
3) $\frac{x}{x^{-5}}$	4) $-\frac{a^{-2}}{b^{-1}}$
5) $\frac{7}{x^{-5}}$	6) $\frac{2b}{-5c^{-2}}$
7) $\frac{2n^{-1}}{12p^{-2}}$	8) $\frac{8b^{-4}}{3c^{-2}}$
9) $89xy^{-2}$	10) $(\frac{1}{3})^{-2}$

Score: ....

Answer Key	
1) $\frac{1}{7}$	2) $-\frac{2}{x^2}$
3) x^5	4) $-\frac{b^1}{a^2}$
5) $7x^5$	6) $-2\frac{bc^2}{5}$
7) $\frac{p^2}{6n}$	8) $\frac{8c^2}{3b^4}$
9) $\frac{89x}{y^2}$	10) 9

Name: ....

Writing Scientific Notation

✓ It is used to write very big or very small values in decimal representation.

✓ In scientific notation all numbers can be written in the form of:

$$m \times 10^n, \text{ where } 1 \leq m \leq 10 \text{ and n is any integer.}$$

Decimal notation	Scientific notation
5	5×10^0
$-25,000$	-2.5×10^4
0.5	5×10^{-1}
2,122.456	2.122456×10^3

EXAMPLE:

Write 8.3×10^{-5} in standard notation.

$10^{-5} \rightarrow$ When the decimal moved to the right, the exponent is negative.

Then: $8.3 \times 10^{-5} = 0.000083$

PRACTICES:

Write each number in scientific notation.

1) 12,000,000	2) 25×10^5
3) 0.0015	4) 54,000
5) 0.0005021	6) 666,012
7) 0.00000076	8) 102,900,000
9) 4,100,000,000	10) 3,600,000

Score: ....

Answer Key	
1) 1.2×10^7	2) 2.5×10^6
3) 1.5×10^{-3}	4) 5.4×10^4
5) 5.021×10^{-4}	6) 6.66012×10^5
7) 7.6×10^{-7}	8) 1.029×10^8
9) 4.1×10^9	10) 3.6×10^6

Name: ....

Square Roots

✓ A square root of x is a number p whose square is: $p^2 = x$

  p is a square root of x.

EXAMPLE:

Find the square root of $\sqrt{225}$.

  First factor the number: $225 = 15^2$, Then: $\sqrt{225} = \sqrt{15^2}$

  Now use radical rule: $\sqrt[n]{a^n} = a$

  Then: $\sqrt{15^2} = 15$

PRACTICES:

Find the value each square root.

1) $\sqrt{25}$	2) $\sqrt{1,600}$
3) $\sqrt{100}$	4) $\sqrt{121}$
5) $\sqrt{4}$	6) $\sqrt{225}$
7) $\sqrt{10,000}$	8) $\sqrt{16}$
9) $\sqrt{64}$	10) $\sqrt{36}$

Score: ....

Answer Key	
1) 5	2) 40
3) 10	4) 11
5) 2	6) 15
7) 100	8) 4
9) 8	10) 6

Chapter 5 : Algebraic Expressions

Topics that you'll learn in this chapter:

- ➢ Expressions and Variables

- ➢ Simplifying Variable and Polynomial Expressions

- ➢ Translate Phrases into an Algebraic Statement

- ➢ The Distributive Property

- ➢ Evaluating One and two Variable

- ➢ Combining like Terms

"Without mathematics, there's nothing you can do. Everything around you are mathematics. Everything around you are numbers." – *Shakuntala Devi*

Name: ....

Translate Phrases into an Algebraic Statement

How to translate key words and phrases into algebraic expressions:

- ✓ Addition: the sum of, more than, plus, etc.
- ✓ Subtraction: less than, decreased, minus, etc.
- ✓ Multiplication: times, multiplied, product, etc.
- ✓ Division: quotient, ratio, divided, etc.

EXAMPLE:

5 times the sum of x and 8

Sum of 8 and x: $8 + x$. Times is used for multiplication. Then: $5 \times (8 + x)$

PRACTICES:

Write an algebraic expression for each phrase.

1) Fifteen subtracted from a number.

2) The quotient of seventeen and a number.

3) A number increased by fifty.

4) A number divided by – 21.

5) The difference between sixty –three and a number.

6) Threefold a number decreased by 45.

7) Seven times the sum of a number and – 21.

8) The quotient of 90 and the product of a number and – 8.

9) Nine subtracted from 4 times a number.

10) The difference of six and a number.

Score: ....

Answer Key	
1) $x - 15$	2) $\frac{17}{x}$
3) $x + 50$	4) $-\frac{x}{21}$
5) $63 - x$	6) $3x - 45$
7) $7(x + (-21))$	8) $-\frac{90}{8x}$
9) $4x - 9$	10) $6 - x$

Name: ....

The Distributive Property

✓ Distributive Property:

$$a(b + c) = ab + ac$$

EXAMPLE:

Simply. $(5x - 3)(-5) =$

Use Distributive Property formula: $a(b + c) = ab + ac$

$$(5x - 3)(-5) = -25x + 15$$

PRACTICES:

Use the distributive property to simply each expression.

1) $4(9 - 3x)$	2) $-(-8 - 4x)$
3) $(-5x - 1)(-2)$	4) $(-3)(2x - 4)$
5) $4(5 + 3x)$	6) $(-9x + 10)3$
7) $(-4 - 5x)(-3)$	8) $(-2x)(-3 + 2x) - 3x(1 - 5x)$
9) $(-2)(3x - 1) + 4(3x + 2)$	10) $(-15)(2x + 3)$

Score: ....

Answer Key	
1) $-12x + 36$	2) $4x + 8$
3) $10x + 2$	4) $-6x + 12$
5) $12x + 20$	6) $-27x + 30$
7) $15x + 12$	8) $11x^2 + 3x$
9) $6x + 10$	10) $-30x - 45$

Name: ....

Evaluating One Variable

✓ To solve a variable expression, find the variable and substitute a number for that variable.
✓ Perform the mathematical operations.

EXAMPLE:

Solve this expression. $12 - 2x$, $x = -1$

First substitute -1 for x, then:

$$12 - 2x = 12 - 2(-1) = 12 + 2 = 14$$

PRACTICES:

Simplify each algebraic expression.

1) $5x + 4, x = 1$	2) $x + (-4), x = -6$
3) $-10x + 8, \ x = -2$	4) $\left(-\frac{36}{x}\right) - 10 + 2x, \ x = 6$
5) $\frac{36}{x} - 3, x = 3$	6) $(-10) - \frac{x}{4} + 4x, \ x = -8$
7) $15 + 6x - 3, x = -1$	8) $(-5) + \frac{x}{8}, \ x = 64$
9) $\left(-\frac{24}{x}\right) - 10 + 5x, x = 4$	10) $(-4) + \frac{4x}{9}, x = 81$

Score: ....

Answer Key	
1) 9	2) −10
3) 28	4) −4
5) 9	6) −40
7) 6	8) 3
9) 4	10) 32

Name: ....

Evaluating Two Variables

To solve an algebraic expression, substitute a number for each variable and perform the mathematical operations.

EXAMPLE:

Solve this expression. $-3x + 5y, x = 2, y = -1$

First substitute 2 for x, and -1 for y, then:

$$-3x + 5y = -3(2) + 5(-1) = -6 - 5 = -11$$

PRACTICES:

Simplify each algebraic expression.

1) $5a - (5 - b)$, $a = 2, b = 3$	2) $5x + 3y - 6 + 3y$, $x = 3, y = 1$
3) $\left(-\frac{27}{x}\right) + 4 + 3y$, $x = 3, y = 5$	4) $(-4)(-3a - 5b)$, $a = 3, b = 4$
5) $7x + 10 - 5y$, $x = 3, y = 6$	6) $18 + 3(-x - 4y)$, $x = 2, y = 5$
7) $12x + 2y$, $x = 5, y = 10$	8) $x \times 6 \div 3y$, $x = 6, y = 1$
9) $4x - 3y$, $x = 6, y = 3$	10) $\left(-\frac{14}{x}\right) + 4y$, $x = 7, y = -3$

Score: ....

Answer Key	
1) 8	2) 15
3) 10	4) 116
5) 1	6) − 48
7) 80	8) 12
9) 15	10) −14

Name: ....

Expressions and Variables

✓ In algebra, a variable is a letter used as a replacement for a number. The most common letters are: $x, y, z, a, b, c, m, and\ n$.

✓ An algebraic expression is an expression that has variables, integers, and math operations such as addition, subtraction, multiplication, division, etc.

✓ In an expression, we can combine "identical" terms. (Values with same power and variable)

EXAMPLE:

Simplify this expression. $(10x + 2x + 3)$ =?

Combine like terms. Then: $(10x + 2x + 3) = 12x + 3$ (remember you cannot combine variables and numbers

PRACTICES:

Simplify each expression.

1) $10(-3 - 8x), x = 4$	2) $-3(5 - 8x) - 6x, x = 1$
3) $2x - 8x, x = 2$	4) $x + 12x, x = 6$
5) $20 - 5x + 10x + 5, x = 3$	6) $15(5x + 3), x = 0$
7) $20(4 - x) - 9, x = 2$	8) $20x - 8x - 10, x = 5$
9) $6x + 9y,\ x = 4, y = 2$	10) $6x - 2x, x = 8$

Score: ....

Answer Key	
1) -350	2) 3
3) -12	4) 78
5) 40	6) 45
7) 31	8) 50
9) 42	10) 32

Name: ....

Combining like Terms

✓ We separate the terms by "+" and "-" signs.

✓ Identical terms are those terms with same powers and same variables. Make sure to use the "+" or "-" that is in front of the coefficient

EXAMPLE:

Simplify this expression. $(-5)(8x - 6) =$

Use Distributive Property formula: $a(b + c) = a + ac$

$(-5)(8x - 6) = -40x + 30$

PRACTICES:

Simplify each expression.

1) $-8(-5x + 1)$	2) $6(-2 + 4x)$
3) $-8 - 14x + 16x + 3$	4) $9x - 7x - 15 + 18$
5) $(-9)(12x - 21) + 31$	6) $2(4x + 9) + 12x$
7) $4(-2x - 17) + 14(3x + 1)$	8) $(9x - 5y)7 + 25y$
9) $4.5x^3 \times (-8x)$	10) $-19 - 15x^2 + 12x^2$

Score: ....

Answer Key	
1) $40x - 8$	2) $24x - 12$
3) $2x - 5$	4) $2x + 3$
5) $220 - 108x$	6) $20x + 18$
7) $34x - 54$	8) $63x - 10y$
9) $-36x^4$	10) $-3x^2 - 19$

Name: ....

Simplifying Polynomial Expressions

✓ A polynomial is a unique expression that consists of coefficients and variables that performs only the arithmetic of addition, subtraction, multiplication, and non-negative integer exponents of variables.

$$P(x) = a_n x^n + a_{n-1} x^{n-1} + \ldots + a_2 x^2 + a_1 x + a_0$$

EXAMPLE:

Simplify this Polynomial Expressions. $4x^2 - 5x^3 + 15x^4 - 12x^3 =$

Combine "like" terms: $-5x^3 - 12x^3 = -17x^3$

Then: $4x^2 - 5x^3 + 15x^4 - 12x^3 = 4x^2 - 17x^3 + 15x^4$

Then write in standard form: $4x^2 - 17x^3 + 15x^4 = 15x^4 - 17x^3 + 4x^2$

PRACTICES:

Simplify each polynomial.

1) $(2x^2 + 4) - (9 + 5x^2)$	2) $(25x^3 - 12x^2) - (6x^2 - 9x^3)$
3) $14x^5 - 15x^6 + 2x^5 - 16x^6 + x^6$	4) $(15 + 12x^3) + (3x^3 + 5)$
5) $13x^3 - 15x^4 + 12x^3 + 20x^4$	6) $-6x^2 + 15x^2 + 17x^3 + 16 - 32$
7) $15x^3 + 12 + 2x^2 - 5x - 10x$	8) $24x^2 - 16x^3 - 4x(2x^2 + 3x)$
9) $(21x^4 - 10x) - (2x - x^4)$	10) $(7x^2 - 9) + (x^2 - 8x^3)$

Score: ....

Answer Key	
1) $-3x^2 - 5$	2) $34x^3 - 18x^2$
3) $-30x^6 + 16x^5$	4) $15x^3 + 20$
5) $5x^4 + 25x^3$	6) $17x^3 + 9x^2 - 16$
7) $15x^3 + 2x^2 - 15x + 12$	8) $-24x^3 + 12x^2$
9) $22x^4 - 12x$	10) $-8x^3 + 8x^2 - 9$

Chapter 6 : Equations and Inequalities

Topics that you'll learn in this chapter:

- ➢ One, Two, and Multi – Step Equations

- ➢ Graphing Single– Variable Inequalities

- ➢ One, Two, and Multi – Step Inequalities

- ➢ Solving Systems of Equations by Substitution and Elimination

- ➢ Finding Slope and Writing Linear Equations

- ➢ Graphing Lines Using Slope– Intercept and Standard Form

- ➢ Graphing Linear Inequalities

- ➢ Finding Midpoint and Distance of Two Points

"The study of mathematics, like the Nile, begins in minuteness but ends in magnificence." –
Charles Caleb Colton

Name: ....

One–Step Equations

✓ The values of two algebraic expressions on both sides of an equation are always equal.

$$ax + b = c$$

✓ You only require performing one Math operation to solve the problem.

EXAMPLE:

Solve this equation. $x + 24 = 0 , x = ?$

Here, we have the addition operation, and its inverse operation is subtraction. To solve this equation, subtract 24 from both sides of the equation: $x + 24 - 24 = 0 - 24$

Then simplify: $x + 24 - 24 = 0 - 24 \rightarrow x = -24$

PRACTICES:

Solve each equation.

1) $x + 4 = 16$	2) $48 = (-2) + x$
3) $5x = (-105)$	4) $(-8) = (8x)$
5) $(-2) = 14 + x$	6) $5 + x = 6$
7) $2x + 3 = (-7)$	8) $28 = x + 7$
9) $(-15) + x = (-15)$	10) $12x = (-36)$

Score: ....

Answer Key

1) 12	2) 50
3) −21	4) −1
5) −16	6) 1
7) −5	8) 21
9) 0	10) −3

Name: ....

Two–Step Equations

- ✓ You only require performing two math operations (add, subtract, multiply, or divide) to solve the equation.
- ✓ Simplify the equation using the inverse of addition or subtraction.
- ✓ Simplify the equation further by using the inverse of division or multiplication.

EXAMPLE:

Solve this equation. $3x = 15, x =?$

Here, we have the multiplication operation (variable x is multiplied by 3) and its inverse operation is division. To solve this problem, we divide both sides of equation by 3:

$3x = 15 \rightarrow 3x \div 3 = 15 \div 3 \rightarrow x = 5$

PRACTICES:

Solve each equation.

1) $4(2 + 2x) = 8$	2) $(-5)(x - 3) = 25$
3) $(-5)(2x - 5) = (-15)$	4) $4(9 + 3x) = -12$
5) $6(2x + 1) = 30$	6) $2(x + 2) = 42$
7) $2(12 + 6x) = 60$	8) $(-10)(5x) = 100$
9) $4(3x + 3) = 24$	10) $\dfrac{x - 5}{3} = 4$

Score: ....

Answer Key	
1) 0	2) −2
3) 4	4) −4
5) 2	6) 10
7) 3	8) −2
9) 1	10) 17

Name: ....

Multi–Step Equations

- ✓ Combine "identical" terms on one side.

- ✓ Put variables to one side by adding or subtracting.

- ✓ Simplify the equation by using the inverse of addition or subtraction.

- ✓ Simplify further by using the inverse of division or multiplication.

EXAMPLE:

Solve this equation. $-(2 - x) = 5$

First, use Distributive Property: $-(2 - x) = -2 + x$

Now by adding 2 to both sides of the equation, we can solve it.

$-2 + x = 5 \rightarrow -2 + x + 2 = 5 + 2$

Now simplify: $-2 + x + 2 = 5 + 2 \rightarrow x = 7$

PRACTICES:

Solve each equation.

1) $8 - 2x = 28$	2) $-10 = -(x + 7)$
3) $2x - 17 = (-x) + 1$	4) $-2x = (-3x) - 8$
5) $5(14 + 2x) + 3x = -x$	6) $x - 11 = x - 5 + 2x$
7) $15 + 2x = (-25) - 2x + 3x$	8) $-3(x - 3x) = 40 - 4x$
9) $24 + 8x + x = (-x + 4)$	10) $-8(1 + 5x) = 152$

Score: ....

Answer Key	
1) −10	2) 3
3) 6	4) −8
5) −5	6) −3
7) −40	8) 4
9) −2	10) −4

Name: ....

Graphing Single–Variable Inequalities

- ✓ Inequality is like equations and uses symbols for "less than" (<) and "greater than" (>).
- ✓ To solve inequalities, we need to separate the variable. (Like in equations)
- ✓ Find the value of the inequality on the number line to graph an inequality.
- ✓ For greater than or less than draw open circle on the value of the variable.
- ✓ Use filled circle if there is an equal sign too.
- ✓ Draw a line to the left or to the right for less or greater than.

EXAMPLE:

Draw a graph for $x > 2$

Since, the variable is greater than 2, then we need to find 2 and draw an open circle above it. Then, draw a line to the right.

PRACTICES:

Draw a graph for each inequality.

1) $2 \geq x$

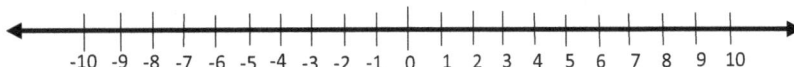

2) $x < 3$

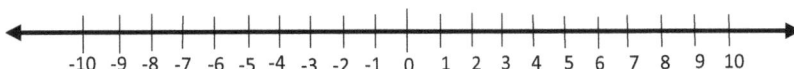

3) $5 \geq x$

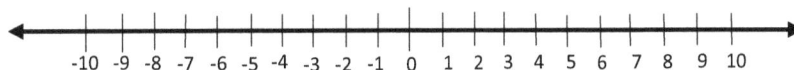

4) $x \geq -2$

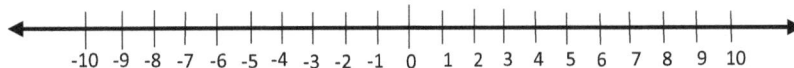

5) $x > 0$

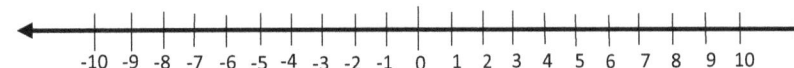

6) $-1.5 < x$

7) $x \geq -1$

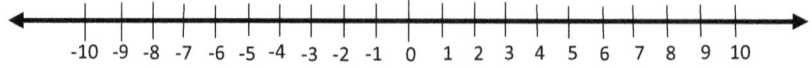

Score: ....

Answer Key

1) $2 \geq x$

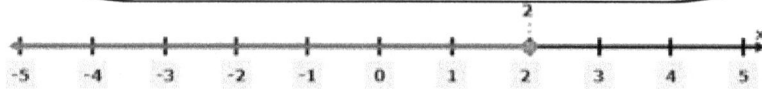

2) $x < 3$

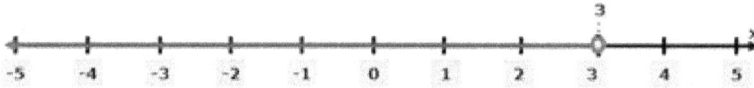

3) $5 \geq x$

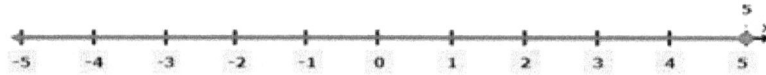

4) $x \geq -2$

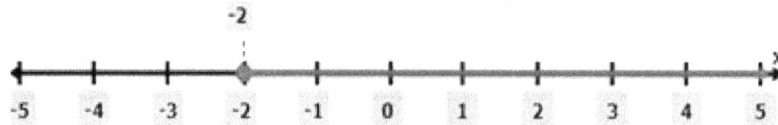

5) $x > 0$

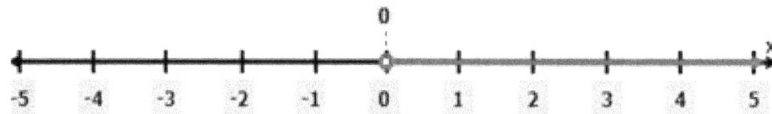

6) $-1.5 < x$

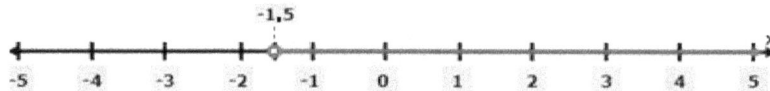

7) $x \geq -1$

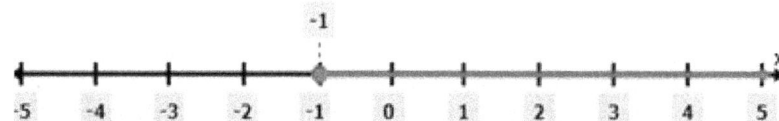

Name: ....

One–Step Inequalities

- ✓ Like equations, first separate the variable by using inverse operation.
- ✓ For dividing or multiplying both sides by negative numbers, flip the direction of the inequality sign.

EXAMPLE:

Solve this inequality. $x - 1 \leq 2$

Add 1 to both sides. $x - 1 \leq 2 \rightarrow x - 1 + 1 \leq 2 + 1$, then: $x \leq 3$

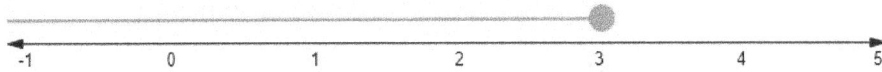

PRACTICES:

Solve each inequality and graph it.

1) $2x + 3 \geq 7$

2) $x < 3$

3) $5 \geq x$

4) $x \geq -2$

5) $x > 0$

6) $-1.5 < x$

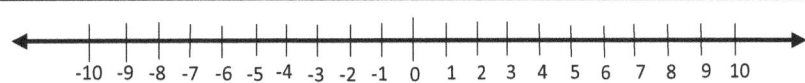

7) $x \geq -1$

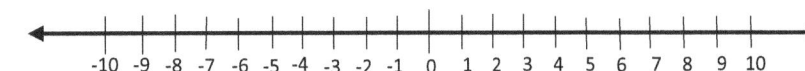

Score: ....

Answer Key

1) $2 \geq x$

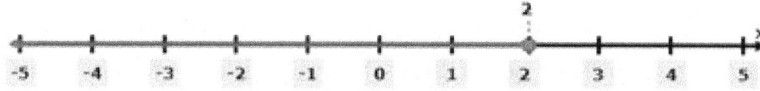

2) $x < 3$

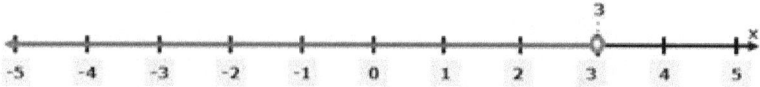

3) $5 \geq x$

4) $x \geq -2$

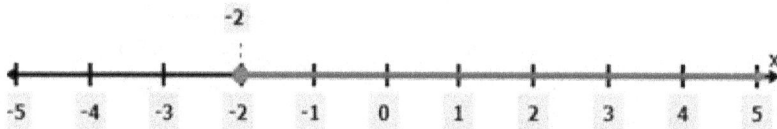

5) $x > 0$

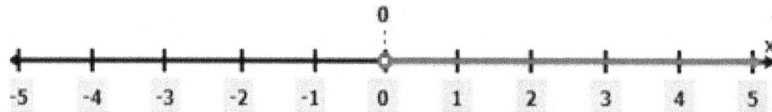

6) $-1.5 < x$

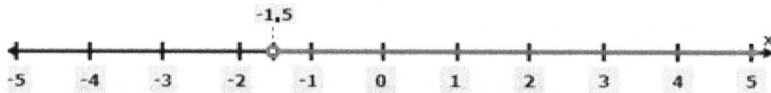

7) $x \geq -1$

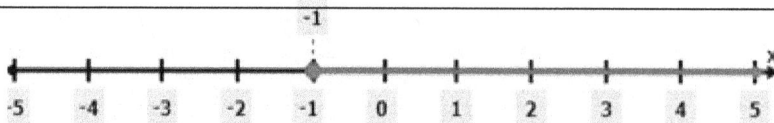

Name: ....

Two–Step Inequalities

- ✓ Separate the variable.

- ✓ Flip the direction of the inequality sign for dividing both sides by negatives numbers.

- ✓ We can simplify by using the inverse of addition or subtraction.

- ✓ We can simplify further by using the inverse of division or multiplication.

EXAMPLE:

Solve: $2x + 9 \geq 11$

First add -9 to both sides: $2x + 9 - 9 \geq 11 - 9 \rightarrow 2x \geq 2$

Now, divide both sides by 2: $2x \geq 2 \rightarrow x \geq 1$

PRACTICES:

Solve each inequality and graph it.

1) $x - 4 \leq 4$	2) $x + 4 \geq 5$
3) $3x - 2 \leq 7$	4) $5x + 2 < 12$
5) $x + 7 \geq 9$	6) $3x - 3 \leq 3$
7) $7x - 4 < 3$	8) $8 + x \leq 13$
9) $2x + 7 \leq 11$	10) $10x - 16 < 4$

Score: ....

Answer Key	
1) $x \le 8$	2) $x \ge 1$
3) $x \le 3$	4) $x < 2$
5) $x \ge 2$	6) $x \le 2$
7) $x < 1$	8) $x \le 5$
9) $x \le 2$	10) $x < 2$

Name: ....

Multi–Step Inequalities

✓ Separate the variable.

✓ We can simplify by using the inverse of addition or subtraction.

✓ We can simplify further by using the inverse of division or multiplication.

EXAMPLE:

Solve this inequality. $2x - 2 \leq 6$

First add 2 to both sides: $2x - 2 + 2 \leq 6 + 2 \rightarrow 2x \leq 8$

Now, divide both sides by 2: $2x \leq 8 \rightarrow x \leq 4$

PRACTICES:

Solve each inequality.

1) $-(x + 3) + 8 < 25$	2) $\dfrac{3x + 1}{2} \leq 5$
3) $\dfrac{x - 4}{3} > 7$	4) $4(x - 2) \leq 8$
5) $\dfrac{x}{3} + \dfrac{1}{3} < 2$	6) $\dfrac{x+4}{5} > 3$
7) $\dfrac{x}{8} + \dfrac{3}{4} < 1$	8) $2(x + 5) + 4 > 10$
9) $24 + 5x < 4$	10) $\dfrac{x+2}{3} > 10$

Score: ....

	Answer Key	
1) $x > -20$		2) $x \leq 3$
3) $x > 25$		4) $x \leq 4$
5) $x < 5$		6) $x > 11$
7) $x < 2$		8) $x > -2$
9) $x < -4$		10) $x > 28$

Name: ....

Solving Systems of Equations by Substitution

✓ Let the system of equations. $x + y = 1; -2x + y = 4$

Put $x = 1 - y$ in the second equation.

$-2(1 - y) + y = 4 \rightarrow -2 + 2y + y = 4 \Rightarrow y = 2$

Put $y = 2$ in $x = 1 - y$; then $x = 1 - 2 = -1 ; (-1,2)$

EXAMPLE:

Solve: $-2x - 2y = -13; -4x + 2y = 10$

For the first equation above, you can add $-4x + 2y$ to the left side and 10 to the right

side of the first equation: $-2x - 2y + (-4x + 2y) = -13 + 10$. Now, if you simplify,

you get: $-2x - 2y - 4x + 2y = -3 \rightarrow -6x = -3 \rightarrow$

$x = 0.5$. Now, put 0.5 for the x in the first equation:

$-2(0.5) - 2y = -13$. By solving this equation, $y = 6$

PRACTICES:

Solve each system of equation by substitution.

1) $-x + 5y = -4$ $x - 3y = 8$	2) $2x + 3y = -6$ $-2x - y = 8$
3) $x + 2y = -5$ $5x - 10y = 5$	4) $y = -x + 5$ $3x - y = -3$
5) $3x = 6$ $10y = 4x + 2$	6) $3x + 2y = 2$ $x + 4y = -6$
7) $4x + y = 3$ $2x + 4y = -2$	8) $4y = 2x + 3$ $x - 4y = -2$
9) $7y = 14x$ $2x - 5y = -24$	10) $5y = x + 2$ $3x - 12y = -5$

Score: ....

Answer Key	
1) $(14, 2)$	2) $(-\frac{9}{2}, 1)$
3) $(-2, -\frac{3}{2})$	4) $(\frac{1}{2}, \frac{9}{2})$
5) $(2, 1)$	6) $(2, -2)$
7) $(1, -1)$	8) $(-1, \frac{1}{4})$
9) $(3, 6)$	10) $(-\frac{1}{3}, \frac{1}{3})$

Name: ....

Solving Systems of Equations by Elimination

- ✓ A system of equations has two equations and two variables. For example, look at the system of equations: $x - y = 1, x + y = 5$
- ✓ The simplest way to solve a system of equation is using the elimination method. The elimination method uses the addition property of equality. On each side of an equation, you can add the same value.

EXAMPLE:

What is the value of x and y in this system of equations? $\begin{cases} 3x - 4y = -20 \\ -x + 2y = 10 \end{cases}$

Solving Systems of Equations by Elimination: $\begin{matrix} 3x - 4y = -20 \\ \underline{-x + 2y = 10} \end{matrix} \Rightarrow$ Multiply the second

equation by 3, then add it to the first equation.

$\begin{matrix} 3x - 4y = -20 \\ \underline{3(-x + 2y = 10)} \end{matrix} \Rightarrow \begin{matrix} 3x - 4y = -20 \\ \underline{-3x + 6y = 30)} \end{matrix} \Rightarrow 2y = 10 \Rightarrow y = 5.$ Now, substitute 5 for y in

the first equation and solve for x. $3x - 4(5) = -20 \rightarrow 3x - 20 = -20 \rightarrow x = 0$

PRACTICES:

Solve each system of equation by elimination.

1) $-5x + y = -5$ $-y = -6x + 6$	2) $-6x - 2y = -2$ $2x - 3y = 8$
3) $5x - 4y = 8$ $-6x + y = -21$	4) $10x - 4y = -24$ $-x - 20y = -18$
5) $25x + 3y = -13$ $12x - 6y = -36$	6) $x - 8y = -7$ $6x + 4y = 10$
7) $-6x + 16y = 4$ $5x + y = 11$	8) $2x - 3y = -10$ $4x + 6y = -20$
9) $x - 5y = -8$ $3x + 7y = -2$	10) $x - 2y = -3$ $2x + 6y = -1$

Score: ....

Answer Key	
1) $(1, 0)$	2) $(1, -2)$
3) $(4, 3)$	4) $(-2, 1)$
5) $(-1, 4)$	6) $(1, 1)$
7) $(2, 1)$	8) $(-5, 0)$
9) $(-3, 1)$	10) $(-2, \frac{1}{2})$

Name: ....

Systems of Equations Word Problems

✓ Define your variables, write the system of equations then use elimination method for solving systems of equations.

EXAMPLE:

Tickets to a movie cost $5 for students and $8 for adults. Some friends purchased **20** tickets for **$115.00**. How many adults ticket did they buy?

Let x be the number of adult tickets and y be the number of student tickets. There are 20 tickets. Then: $x + y = 20$. The cost of students' ticket is $5 and for adults it is $8, and the total cost is $115. So, $8x + 5y = 20$. Now, we have a system of equations:

$$\begin{cases} x + y = 20 \\ 8x + 5y = 115 \end{cases}$$

Multiply the first equation by -5 and add to the second equation: $-5(x + y = 20) =$

$-5x - 5y = -100$

$8x + 5y + (-5x - 5y) = 115 - 100 \rightarrow 3x = 15 \rightarrow x = 5 \rightarrow 5 + y = 20 \rightarrow y = 15.$

There are 5 adults' tickets and 15 student tickets.

PRACTICES:

Solve.

1) A school of 220 students went on a field trip. They took 20 vehicles, some vans, and some minibuses. Find the number of vans and the number of minibuses they took if each van holds 5 students and each minibus hold 15 students.

2) The sum of two numbers is 28. Their difference is 12. Find the numbers.

3) A farmhouse shelters 20 animals, some are pigs, and some are gooses. Altogether there are 64 legs. How many of each animal are there?

4) The sum of the digits of a certain two–digit number is 15. Reversing it's increasing the number by 9. What is the number?

5)	The difference of two numbers is 16. Their sum is 32. Find the numbers.
6)	Tickets to a movie cost $5 for adults and $3 for students. A group of friends purchased 18 tickets for $82.00. How many adults ticket did they buy?
7)	At a store, Eva bought two shirts and five hats for $154.00. Nicole bought three same shirts and four same hats for $168.00. What is the price of each shirt?
8)	A farmhouse shelters 10 animals, some are pigs, and some are ducks. Altogether there are 36 legs. How many pigs are there?
9)	A class of 195 students went on a field trip. They took 19 vehicles, some cars and some buses. If each car holds 5 students and each bus hold 25 students, how many buses did they take?
10)	The sum of two numbers is 50. Their difference is 20. Find the numbers.

Score: ....

Answer Key

1) There are 8 van and 12 minibuses.	2) 8 and 20
3) There are 12 pigs and 8 gooses.	4) 78
5) 24 and 8.	6) 14
7) $32	8) 8
9) 5	10) 15 and 35

Name: ....

Linear Equations

- ✓ The equation of a line: $y = mx + b$

- ✓ Find the slope.

- ✓ Find the y-intercept. This can be done by substituting the slope and the coordinates of a point (x, y) on the line.

EXAMPLE:

What will be the equation of the line that passes through $(2, -2)$ and has a slope of 7?

$y = mx + b$ is the general slope-intercept form of the equation; where, m is the slope and b is the y-intercept.

By substitution of the given point and given slope, we have: $-2 = (2)(7) + b$

So, $b = -2 - 14 = -16$, and our required equation is $y = 7x - 16$.

PRACTICES:

Find the slope of the line through each pair of points.	Write the slope–intercept form of the equation of the line through the given points.
1) $(3, 1), (2, 4)$	2) Through: $(2, 3), (4, 2)$
3) $(-3, 4), (-1, 6)$	4) Through: $(8, -3), (6, 7)$
5) $(4, 4), (6, -6)$	6) Through: $(0.5, 4), (2.5, 4.4)$
7) $(-1, 8), (5, -4)$	8) Through: $(4, -2), (2.5, 1)$
9) $(12, -3), (7, -3)$	10) Through: $(-1, 0.7), (-2.3, 2)$

Score: ....

Answer Key	
1) -3	2) $y = -\frac{1}{2}x + 4$
3) 1	4) $y = -5x + 37$
5) -5	6) $y = \frac{1}{5}x + \frac{39}{10}$
7) -2	8) $y = -2x + 6$
9) 0	10) $y = -x - 0.3$

Name: ....

Graphing Lines of Equations

✓ The equation of the line is: $y = mx + c$

If slope-intercept form of a line given the slope m and the y-intercept (the intersection of the line and y-axis)

EXAMPLE:

Sketch the graph of $y = 8x - 3$.

To graph this line, two points are required.

We have value of y is -3 when x is 0.

The value of x is 3/8 when y is 0.

$x = 0 \rightarrow y = 8(0) - 3 = -3,$

$y = 0 \rightarrow 0 = 8x - 3 \rightarrow x = \dfrac{3}{8}$

Now, the two points are: $(0, -3)$ and $(\frac{3}{8}, 0)$.

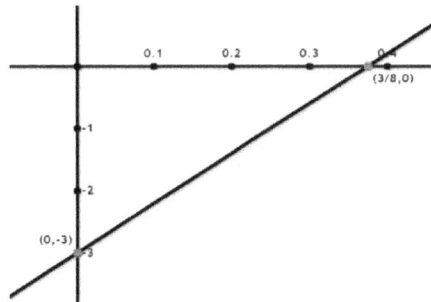

Find the points and graph the line. The slope of the line is 8.

PRACTICES:

Sketch the graph of each line.

1) $y = 3x - 2$

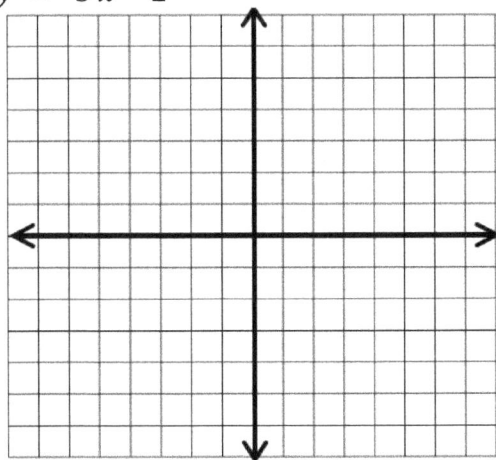

2) $y = 2x + 3$

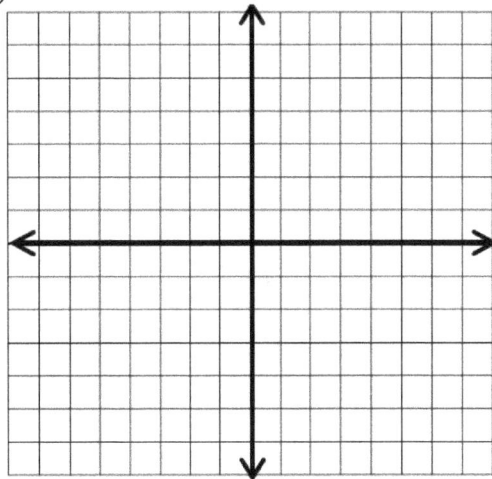

3) $-2x = y + 5$

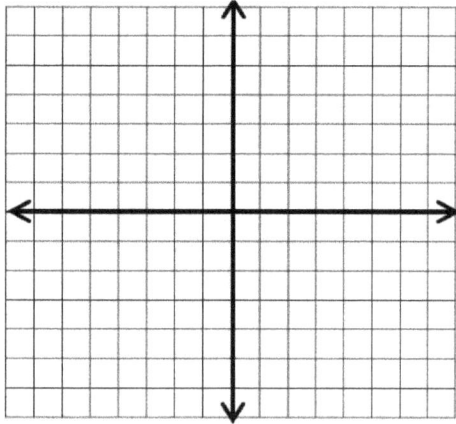

4) $4x + y = 2$

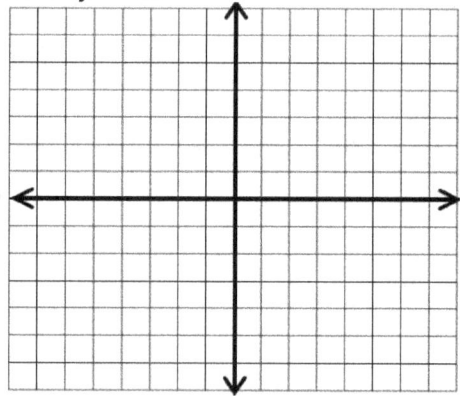

Score: ....

Answer Key

1)

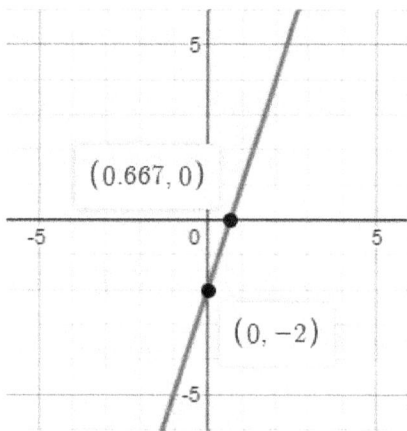

$(0.667, 0)$

$(0, -2)$

2)

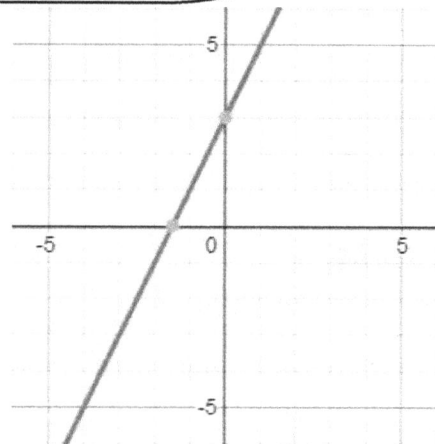

3)

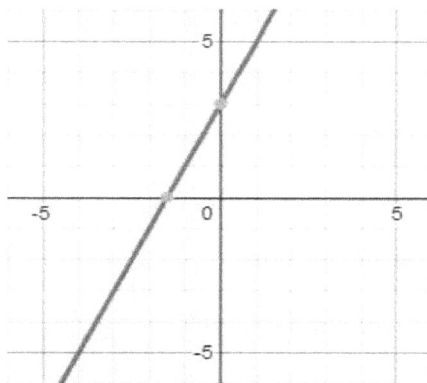

4)

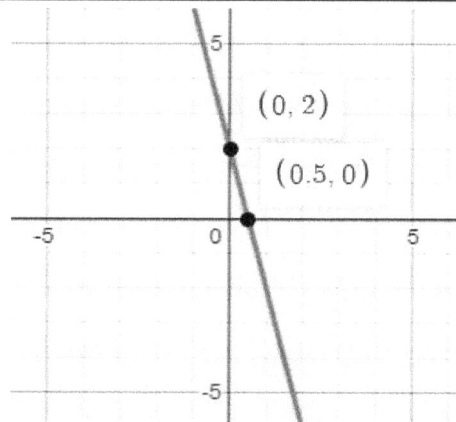

$(0, 2)$

$(0.5, 0)$

Name: ....

Graphing Linear Inequalities

- ✓ First step is to graph the "equals" line.
- ✓ Choose a testing point. (It can be any point on both sides of the line.)
- ✓ Put the value of (x, y) of that point in the inequality. If this satisfy the inequality, then this part of line is solution. If it does not satisfy then other part of line, is solution.

EXAMPLE:

Plot the graph of $y < 2x - 3$.

We know that first step is to graph the line: $y = 2x - 3$.

Now y-intercept is -3 and slope is 2. Then, select a

testing point. The simplest point to test is the origin: $(0, 0)$

$(0,0) \rightarrow y < 2x - 3 \rightarrow 0 < 2(0) - 3 \rightarrow 0 < -3$

0 is greater than -3. So, the other part of the line

(On the right side) is the solution.

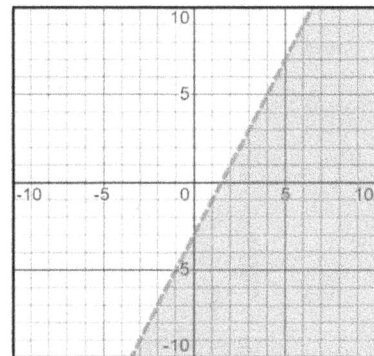

PRACTICES:

Sketch the graph of each linear inequality.

1) $2y + 8x \geq 4$

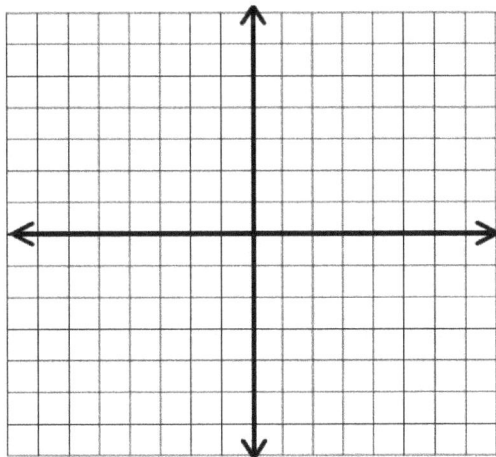

2) $-x + 2y \leq 4$

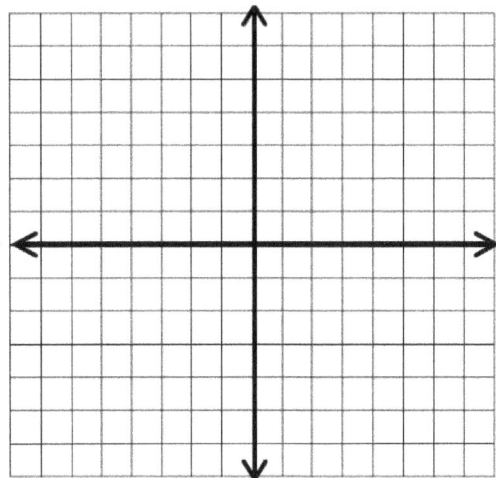

3) $2x + \frac{1}{2}y < 2$

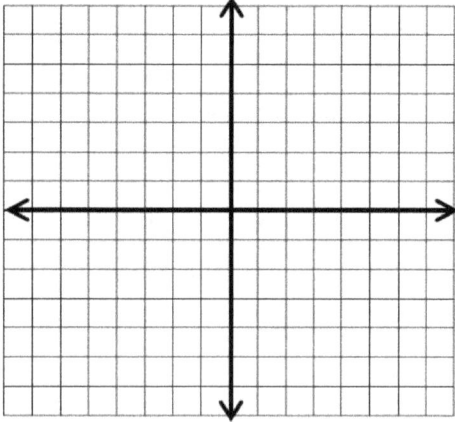

4) $-\frac{1}{3}x + y < 2$

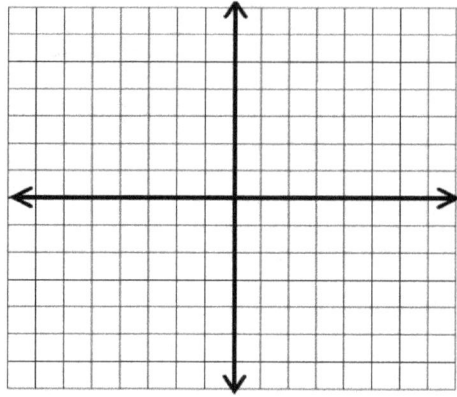

Score: ....

Answer Key

1)

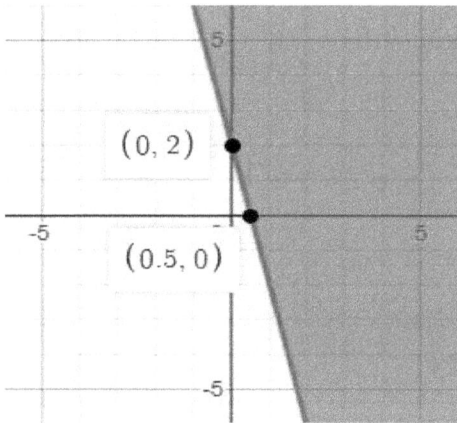

2)

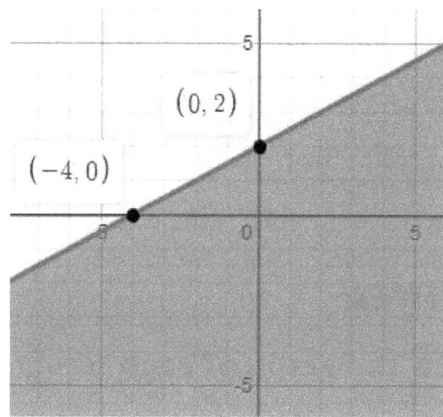

3)

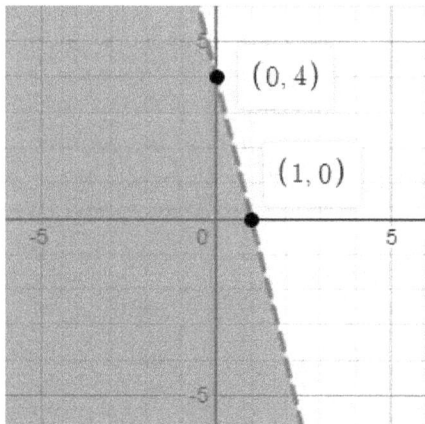

4)

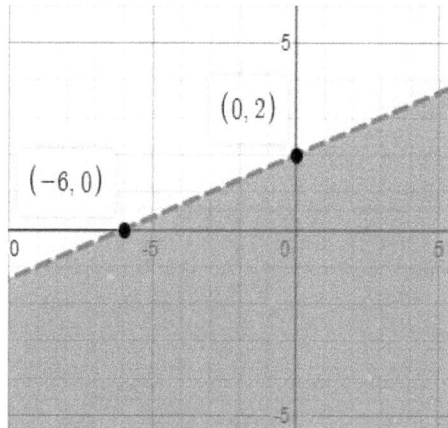

Name: ....

Finding Distance of Two Points

- ✓ Distance of wo points A (x_1, y_1) and B (x_2, y_2): $d = \sqrt{(x_1 - x_2)^2 + (y_1 - y_2)^2}$

- ✓ The midpoint is middle of a line segment.

- ✓ We can find Midpoint of two endpoints A (x_1, y_1) and B (x_2, y_2) using this formula: $M\left(\frac{x_1+x_2}{2}, \frac{y_1+y_2}{2}\right)$

EXAMPLE:

Find the distance between of $(0, 8)$, $(-4, 5)$.

Use distance of two points formula: $d = \sqrt{(x_1 - x_2)^2 + (y_1 - y_2)^2}$

$(x_1, y_1) = (0, 8)$ and $(x_2, y_2) = (-4, 5)$. Then: $d = \sqrt{(x_1 - x_2)^2 + (y_1 - y_2)^2} \rightarrow$

$d = \sqrt{(0 - (-4))^2 + (8 - 5)^2} = \sqrt{(4)^2 + (3)^2} = \sqrt{16 + 9} = \sqrt{25} = 5 \rightarrow d = 5$

PRACTICES:

Find the midpoint of the line segment with the given endpoints.	Find the distance between each pair of points.
1) $(1.5, -1), (0.5, -1)$	2) $(3, 4), (2, -1)$
3) $(1.5, -1), (0.5, -1)$	4) $(6, -1), (2, 3)$
5) $(0, 3), (4, -9)$	6) $(2, 5), (-2, 5)$
7) $(5, 2), (1, 5)$	8) $(0, -4), (-5, 1)$
9) $(-2, 0), (3, -4)$	10) $(3, -2), (-1, -5)$

Score: ....

	Answer Key	
1) $(1,-1)$	2) 5.09
3) $(1.5,-2)$	4) 5.656
5) $(2,-3)$	6) 4
7) $(3,3.5)$	8) 7.07
9) $(0.5,-2)$	10) 5

Chapter 7 : Polynomials

Topics that you'll learn in this chapter:

- ➤ Classifying Polynomials

- ➤ Writing Polynomials in Standard Form

- ➤ Simplifying Polynomials

- ➤ Adding and Subtracting Polynomials

- ➤ Multiplying and Dividing Monomials

- ➤ Multiplying a Polynomial and a Monomial

- ➤ Multiplying Binomials

- ➤ Factoring Trinomials

- ➤ Operations with Polynomials

"Mathematics – the unshaken Foundation of Sciences, and the plentiful Fountain of Advantage to human affairs." — Isaac Barrow

Name: ....

Classifying Polynomials

Name	Degree	Example
constant	0	4
linear	1	$2x$
quadratic	2	$x^2 + 5x + 6$
cubic	3	$x^3 - x^2 + 4x + 8$
quartic	4	$x^4 + 3x^3 - x^2 + 2x + 6$
quantic	5	$x^5 - 2x^4 + x^3 - x^2 + x + 10$

EXAMPLE:

$17x^5 \Rightarrow$ Quantic binomial

PRACTICES:

Name each polynomial by degree and number of terms.	Write each polynomial in standard form
1) -5	2) $12x^4 + x - 4x^3$
3) $x + 1$	4) $12 - x^3 - 3x^5 + 9x^4$
5) $8x^6 - 7$	6) $x^2 + 13x^5 + x^3 - 4x$
7) $3x^2 - x$	8) $x^5 + 2x^3 (x^2 + 2)$
9) $-8x^4 + 3x^3 - 2x^2 - 3x$	10) $(x - 5)(x + 5)$

Score: ....

Answer Key	
1) Constant monomial	2) $12x^4 - 4x^3 + x$
3) Linear binomial	4) $-3x^5 + 9x^4 - x^3 + 12$
5) Sixth degree binomial	6) $13x^5 + x^3 + x^2 - 4x$
7) Quadratic binomial	8) $2x^6 + x^5 + 4x^3$
9) Quartic polynomial with four terms	10) $x^2 - 25$

Adding and Subtracting Polynomials

- ✓ To add polynomials, we combine like terms with some order of operations considerations thrown in.
- ✓ You have to be careful with a negative sign and don't confuse it with addition and subtraction.

EXAMPLE:

Add expressions. $(2x^3 - 6) + (9x^3 - 4x^2) = ?$

Remove parentheses: $(2x^3 - 6) + (9x^3 - 4x^2) = 2x^3 - 6 + 9x^3 - 4x^2$

Now combine like terms: $2x^3 - 6 + 9x^3 - 4x^2 = 11x^3 - 4x^2 - 6$

PRACTICES:

Simplify each expression.

1) $(x^3 + 6) - (6 + 3x^3)$	2) $(x^2 + 8) + (7x^2 - 8)$
3) $(2x^2 + x^3) - (5x^2 + 1)$	4) $(6x^2 - 4x) + (3x - 6x^2 + 1)$
5) $(x - 2x^3) - (4x^3 + 4)$	6) $(2x^3 + 2x^2) - (2x^2 - x^3 + 2)$
7) $(4x^2 - 3) + (x^2 - x^3)$	8) $(x^3 + 13x^4) - (13x^4 + 3x^3)$
9) $(x^4 + 2x^5 + 3x^3) + (4x^3 + 6x^4)$	10) $(3x^3 - 6x^6) + (3x^3 + 2x^6)$

Score: ....

Answer Key

1) $-2x^3$	2) $8x^2$
3) $x^3 - 3x^2 + 1$	4) $-x + 1$
5) $-6x^3 + x - 4$	6) $3x^3 - 2$
7) $-x^3 + 5x^2 - 3$	8) $-2x^3$
9) $2x^5 + 7x^4 + 7x^3$	10) $-4x^6 + 6x^3$

Name: ....

Multiply and Divide Monomials

✓ When dividing monomials, we divide coefficients first and then divide their variables.
✓ If we have exponents with the same base, we'll subtract their powers.
✓ Exponent's rules:

$$x^a \times x^b = x^{a+b}, \qquad \frac{x^a}{x^b} = x^{a-b}$$

$$\frac{1}{x^b} = x^{-b}, \qquad (x^a)^b = x^{a \times b}$$

$$(xy)^a = x^a \times y^a$$

EXAMPLE:

Multiply expressions. $(-3x^7)(4x^3) =?$

Use this formula: $x^a \times x^b = x^{a+b} \rightarrow x^7 \times x^3 = x^{10}$; Then: $(-3x^7)(4x^3) = -12x^{10}$

Dividing expressions. $\frac{18x^2y^5}{2xy^4} =?$

Use this formula: $\frac{x^a}{x^b} = x^{a-b}$, $\frac{x^2}{x} = x^{2-1} = x$ and $\frac{y^5}{y^4} = y^{5-4} = y$; Then: $\frac{18x^2y^5}{2xy^4} = 9xy$

PRACTICES:

Simplify.

1) $(x^3y^2)(42y^4)$	2) $\frac{100x^5y^6}{25x^6y^{11}}$
3) $(8x^4)(12x^5)$	4) $\frac{75x^{16}y^{10}}{5x^6y^7}$
5) $(-2x^{-3}y^2)^2$	6) $\frac{15x^{12}y^5}{5x^9y^2}$
7) $(11x^2y^4)(4x^9y^{10})$	8) $\frac{50x^4y^7}{25x^3y^7}$
9) $(2x^{-3}y^4)^2$	10) $\frac{-21x^8y^{13}}{3x^6y^6}$

Score: ....

Answer Key	
1) $42x^3y^6$	2) $4x^{-1}y^{-5}$
3) $96x^9$	4) $15x^{10}y^3$
5) $4x^{-6}y^4$	6) $3x^3y^3$
7) $44x^{11}y^{14}$	8) $2x$
9) $4x^{-6}y^8$	10) $-7x^2y^7$

Name: ....

Multiplying Monomials

✓ A polynomial having only one term is called monomial, like $2x$ or $7y$

EXAMPLE:

Multiply expressions. $5a^4b^3 \times 2a^3b^2 = ?$

Use this formula: $x^a \times x^b = x^{a+b}$

$a^4 \times a^3 = a^{4+3} = a^7$ and $b^3 \times b^2 = b^{3+2} = b^5$

Then: $5a^4b^3 \times 2a^3b^2 = 10a^7b^5$

PRACTICES:

Simplify each expression.

1) $2xy^2 \times 3z^2$	2) $3xyz \times 5x^2y$
3) $4pq^3 \times (-3p^3q)$	4) $s^3t^2 \times 2st^5$
5) $5p^3 \times (-2p^2)$	6) $-2p^2r \times 6pr^3$
7) $(-a)(-4a^6b)$	8) $2u^2v^3 \times (-8u^3v^3)$
9) $6u^3 \times (2u)$	10) $-5y^2 \times 4x^2y$

Score: ....

Answer Key	
1) $6xy^2z^2$	2) $15x^3y^2z$
3) $-12p^4q^4$	4) $2s^4t^{10}$
5) $-10p^5$	6) $-12p^3r^4$
7) $4a^7b$	8) $-16u^5v^6$
9) $12u^4$	10) $-20x^2y^3$

Name: ....

Multiply a Polynomial and a Monomial

- ✓ Use the product rule for exponents to multiply monomials.
- ✓ Use distributive property to multiply a monomial by a polynomial.

$$a \times (b + c) = a \times b + a \times c$$

EXAMPLE:

Multiply expressions. $-4x(5x + 9) = ?$

Use Distributive Property: $-4x(5x + 9) = -20x^2 - 36x$

PRACTICES:

Find each product.

1) $3(2x - 2y)$	2) $5x(4x - y)$
3) $-2x(x + 5)$	4) $11(3x + 7)$
5) $10x(5x - 2y)$	6) $4(3x - 5y)$
7) $2x(3x^3 - 5x + 4)$	8) $-4x (2 + 4xy)$
9) $3(2x^2 - 8x + 3)$	10) $-3x^2(3x^2 + 5)$

Score: ....

Answer Key	
1) $6x - 6y$	2) $20x^2 - 5xy$
3) $-2x^2 - 10$	4) $33x + 77$
5) $50x^2 - 20xy$	6) $50x^2 - 20xy$
7) $6x^4 - 10x^2 + 8x$	8) $-16x^2y - 8x$
9) $6x^2 - 24x + 9$	10) $-9x^4 - 15x^2$

Name: ....

Multiply Binomials

✓ Use "FOIL". (First-Out-In-Last)

$$(x + a)(x + b) = x^2 + (b + a)x + ab$$

EXAMPLE:

Multiply Binomials. $(x + 5)(x - 2) =?$

Use "FOIL". (First–Out–In–Last):

$(x + 5)(x - 2) = x^2 - 2x + 5x - 10$
Then simplify: $x^2 - 2x + 5x - 10 = x^2 + 3x - 10$

PRACTICES:

Multiply.

1) $(2x - 2)(x + 3)$	2) $(4x + 2)(2x + 1)$
3) $(x + 3)(x + 4)$	4) $(x^2 + 5)(x^2 - 5)$
5) $(2x - 3)(x + 4)$	6) $(2x - 6)(x + 7)$
7) $(x - 2)(3x - 4)$	8) $(2x - 5)(x + 4)$
9) $(x + 10)(x - 10)$	10) $(x - 3)(3x + 4)$

Score: ....

Answer Key	
1) $2x^2 + 4x - 6$	2) $8x^2 + 8x + 2$
3) $x^2 + 7x + 12$	4) $x^4 - 25$
5) $2x^2 + 5x - 12$	6) $2x^2 + 8x - 42$
7) $3x^2 - 10x + 8$	8) $2x^2 + 3x - 20$
9) $x^2 - 100$	10) $3x^2 - 5x - 12$

Name: ....

Factor Trinomials

✓ FOIL":
$$(x + a)(x + b) = x^2 + (b + a)x + ab$$

✓ "Difference of Squares":
$$a^2 - b^2 = (a + b)(a - b)$$
$$a^2 + 2ab + b^2 = (a + b)(a + b)$$
$$a^2 - 2ab + b^2 = (a - b)(a - b)$$

✓ "Reverse FOIL":
$$x^2 + (b + a)x + ab = (x + a)(x + b)$$

EXAMPLE:

Factor this trinomial. $x^2 - 2x - 8 =$?

Expression is broken into groups: $(x^2 + 2x) + (-4x - 8)$

Now x is a factor from $x^2 + 2x$: $x(x + 2)$ and factor out -4 from $-4x - 8$: $-4(x + 2)$

Then: $= x(x + 2) - 4(x + 2)$, now factor out like term: $x + 2$

Then: $(x + 2)(x - 4)$

PRACTICES:

Factor each trinomial.

1) $x^2 - 12x + 27$	2) $x^2 + 5x - 24$
3) $x^2 + 13x + 30$	4) $x^2 - 81$
5) $2x^2 + 12x - 14$	6) $x^2 + 2x - 8$
7) $2x^2 + 3x + 1$	8) $2x^2 + 2x - 4$
9) $9x^2 + 3x - 2$	10) $x^2 + 15x + 56$

Score: ....

Answer Key	
1) $(x - 3)(x - 9)$	2) $(x + 8)(x - 3)$
3) $(x + 10)(x + 3)$	4) $(x + 9)(x - 9)$
5) $(x + 7)(2x - 2)$	6) $(x - 2)(x + 4)$
7) $(2x + 1)(x + 1)$	8) $(2x - 2)(x + 2)$
9) $(3x - 1)(3x + 2)$	10) $(x + 7)(x + 8)$

Name: ....

Operations with Polynomials

✓ Use distributive property to multiply a monomial by a polynomial.

$$a \times (b + c) = a \times b + a \times c$$

EXAMPLE:

Multiply. $5(2x - 6) =$

Use the distributive property: $5(2x - 6) = 5 \times 2x - 5 \times (-6) = 10x - 30$

PRACTICES:

Find each product.

1) $x^2(3x - 2)$	2) $2x^2(5x - 3)$
3) $-x(5x - 3)$	4) $x^2(-3x + 9)$
5) $5(7x + 3)$	6) $8(3x + 8)$
7) $5(10x + 4)$	8) $-3x^5(x - 3)$
9) $5(3x^2 - x + 2)$	10) $4(x^2 - 2x + 3)$

Score: ....

Answer Key	
1) $3x^3 - 2x^2$	2) $10x^3 - 6x^2$
3) $-5x^2 + 3x$	4) $-3x^3 + 9x^2$
5) $35x + 15$	6) $24x + 64$
7) $50x + 20$	8) $-3x^6 + 9x^5$
9) $15x^2 - 5x + 10$	10) $4x^2 - 8x + 12$

Name: ....

Simplifying Polynomials

✓ Find "identical" terms. (They have same variables with same power).

✓ Use "FOIL". (First-Out-In-Last) for binomials:

$$(x + a)(x + b) = x^2 + (b + a)x + ab$$

✓ Using order of operation, add or subtract "identical" terms

EXAMPLE:

Simplify this expression. $(4 + x)(x - 3) =$?

Use FOIL: $(x + 4)(x - 3) = x^2 + x - 12$

PRACTICES:

Simplify each expression.

1) $-3x^2 + x^5 + 7x^5 - 2x^2 + 6$	2) $18x^5 - 3x^5 + 7x^2 - 15x^5 + 4$
3) $x(x^3 + 9) - 6(8 + x^2)$	4) $x(x^2 + 2x^3) - x^3 + x$
5) $4 - 17x^2 + 30x^2 - 17x^2 + 26$	6) $4x^2 - 8x + 3x^3 + 15x - 20x$
7) $(x - 6)(x - x^2 + 5)$	8) $(x - 5)(x + 5)$
9) $(x^4 - x) + (4x^2 - 3x^4)$	10) $x(x^2 + x + 3)$

Score: ....

Answer Key	
1) $8x^5 - 5x^2 + 6$	2) $7x^2 + 4$
3) $x^4 - 6x^2 + 9x - 48$	4) $2x^4 + x$
5) $-4x^2 + 30$	6) $3x^3 + 4x^2 - 13x$
7) $-x^3 + 7x^2 - x - 30$	8) $x^2 - 25$
9) $-2x^4 + 4x^2 - x$	10) $x^3 + x^2 + 3x$

Chapter 8 : Functions

Topics that you'll learn in this chapter:

- ➢ Relations and Functions
- ➢ Rate of change and Slope
- ➢ x and y intercept
- ➢ Slope-intercept form
- ➢ Slope-point form
- ➢ Equation of Parallel or Perpendicular lines
- ➢ Equation of Horizontal and Vertical Lines
- ➢ Function Notation
- ➢ Adding and Subtracting Functions
- ➢ Multiplying and Dividing Functions
- ➢ Composition of Functions
- ➢ Solve a Quadratic Equation

It's fine to work on any problem, so long as it generates interesting mathematics along the way – even if you don't solve it at the end of the day." – Andrew Wiles

Name:

....

Relations and Functions

✓ **RELATION**: A relationship between two sets of elements like input and output, input can have as many as outputs.

✓ **FUNCTION**: A relationship between two sets of elements like input and output and only and exactly one output related to one input.

✓ A function is a type of relation, but a relation may not be a function.

✓ **The Vertical Line Test**

A graphical method that we can observe the cross between the graph and vertical line.

The graph is a function if and only if the intersection point is one.

EXAMPLE:

Use the vertical line test to determine if the graph is function or not?

The graph is not a function

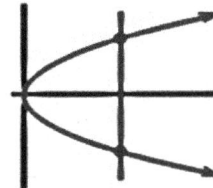

PRACTICES:

State the domain and range of each relation. Then determine whether each relation is a function.

1)

Function:

....

Domain:

....

Range:

....

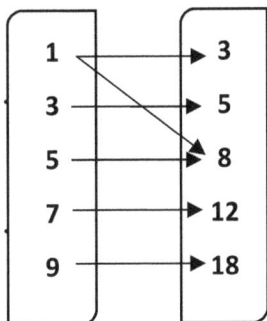

2)

Function:

....

Domain:

....

Range:

....

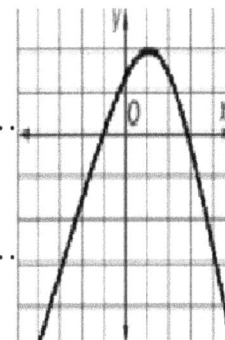

3) $\{(1,-2),(4,-1),(0,5),(4,0),(3,8)\}$ Function: ………………… Domain: ………………… Range: …………………	4) Function: ………………… Domain: ………………… Range: …………………

x	y
3	4
0	1
−2	−3
6	−3
8	2

Score: …………………………………

Answer Key

1) No, $D_f = \{1,3,5,7,9\}$, $R_f = \{3,5,8,12,18\}$	2) Yes, $D_f = (-\infty,\infty)$, $R_f = \{2,-\infty)$
3) No, $D_f = \{1,4,0,3\}$, $R_f = \{-2,-1,5,0,8\}$	4) Yes, $D_f = \{3,0,-2,6,8\}$, $R_f = \{4,1,-3,2\}$

Name: ....

Rate of change

- ✓ Slope can be described as "rate of change".

- ✓ Rate of change is a ratio between a change in one variable comparing to a corresponding change in another variable. Rate of change $= \frac{change\ in\ output\ (y)}{change\ in\ input\ (x)}$

- ✓ Rates of change can be positive, negative, or zero.

EXAMPLE:

The table shows the amount of money SB carwash made washing car. Find the rate of change in dollar per car?

SB Carwash	Number	4	8	12	16
	Money ($)	32	56	80	104

Rate of change $= \frac{change\ in\ output\ (y)}{change\ in\ input\ (x)} = \frac{Change\ in\ money}{Change\ in\ car} = \frac{56-32}{8-4} = \frac{24}{4} = \frac{6}{1}$, or \$6 per car

PRACTICES:

What is the average rate of change of the function?

1)

Gallons	3	5	7	9
Miles	81	135	189	243

2)

Products	145	159	173	187
Costs	761	719	677	635

3)

x	4.5	6	7.5	9
y	6	15	24	33

4)

x	41	47	53	59
y	67	52	37	22

5) $f(x) = -2x + 4$, from $x = -1$ to $x = 4$?

6) $f(x) = x - 6$, from $x = -5$ to $x = 1$?

7) $f(x) = -4$, from $x = 3$ to $x = -2$?

8) $f(x) = 3x^2 + 5$, from $x = 3$ to $x = 6$?

9) $f(x) = -2x^2 - 4$, from $x = 2$ to $x = 4$?

10) $f(x) = x^3 + 3$, from $x = 1$ to $x = 2$?

Score: ....

Answer Key	
1) 27 miles per gallon	2) −3 cost per product
3) 6	4) −2.5
5) −2	6) 1
7) 0	8) 27
9) −12	10) 7

Name: ....

Slope

✓ The slope is used to describe the steepness and direction of lines on the coordinate plane.

✓ A coordinate plane is a two-dimensional plane formed by the intersection of a vertical line called y-axis and a horizontal line called x-axis. These are perpendicular lines that intersect each other at zero, and this point is called the origin.

✓ An ordered pair (x, y) shows the location of a point.

✓ A line on coordinate plane can be drawn by connecting two points.

✓ The slope of a line with two points A (x_1, y_1) and B (x_2, y_2) can be found by using this formula: $\frac{y_2 - y_1}{x_2 - x_1} = \frac{rise}{run}$

EXAMPLE:

Use the given points to determine the slope. Points, $(3, -7), (-2, -9)$

$$\text{Slope} = \frac{y_2 - y_1}{x_2 - x_1} = \frac{-9-(-7)}{-2+3} = \frac{-9+7}{1} = \frac{-2}{1} = -2$$

PRACTICES:

Find the slope of the line through each pair of points.

1) $(2, -9), (5, -6)$	2) $(14, 7), (20, 12)$
3) $(1, -5), (8, -4)$	4) $(13, -9), (15, -7)$
5) $(-5, -8), (-8, -2)$	6) $(0, 0), (12, -2)$
7) $(14, -8), (-6, 5)$	8) $(-2, 5), (-2, 8)$
9) $(-14, -9), (-6, -15)$	10) $(-19, 2), (2, -19)$

Score: ....

	Answer Key	
1) 1	2) $\frac{5}{6}$
3) $\frac{1}{7}$	4) 1
5) -2	6) $-\frac{1}{6}$
7) $-\frac{13}{20}$	8) Undefined
9) $-\frac{3}{4}$	10) -1

Name: ....

x and y intercept

- ✓ *x*-intercept is the point at which the graph crosses the x-axis, and the value of y is zero.
- ✓ *y*-intercept is the point at which the graph crosses the y-axis, and the value of x is zero.

EXAMPLE:

Find the intercepts of the equation $y = 3x - 21$.

To find the *x*-intercept, set $y = 0$, then $0 = 3x - 21 \rightarrow 3x = 21 \rightarrow x = 7$. *x*-intercept(7,0)

To find the *y*-intercept, set $y = 0$, then $y = 3(0) - 21 \rightarrow y = -21$. y-intercept $(0, -12)$

PRACTICES:

Find the x and y intercepts for the following equations.

1) $5x + 3y = 15$	2) $y = x + 8$
3) $4x = y + 16$	4) $x + y = -2$
5) $4x - 3y = 7$	6) $7y - 5x + 10 = 0$
7) $\frac{3}{7}x + \frac{1}{4}y + \frac{2}{3} = 0$	8) $3x - 21 = 0$
9) $24 - 4y = 0$	10) $-2x - 6y + 42 = 12$

Score: ....

Answer Key

1) $y - intercept = 5$
   $x - intercept = 3$

2) $y - intercept = 8$
   $x - intercept = -8$

3) $y - intercept = -16$
   $x - intercept = 4$

4) $y - intercept = -2$
   $x - intercept = -2$

5) $y - intercept = -\frac{7}{3}$
   $x - intercept = \frac{7}{4}$

6) $y - intercept = -\frac{10}{7}$
   $x - intercept = 2$

7) $y - intercept = -\frac{8}{3}$
   $x - intercept = -\frac{2}{7}$

8) $y - intercept = undefind$
   $x - intercept = 7$

9) $y - intercept = 6$
   $x - intercept = undefind$

10) $y - intercept = 5$
    $x - intercept = 15$

Name: ....

Writing Linear Equations

✓ The equation of a line:

$$y = mx + b$$

✓ Identify the slope.

✓ Find the y-intercept. This can be done by substituting the slope and the coordinates of a point (x, y) on the line.

EXAMPLE:

If f is a linear function, with $f(4) = -2$, and $f(7) = 1$, find an equation for the function.

Slope: $\frac{y_2 - y_1}{x_2 - x_1} = \frac{1-(-2)}{7-4} = \frac{1+2}{3} = 1$

$y = mx + b \rightarrow y = 1 \times x + b \rightarrow y = x + b$, and $f(7) = 1$, then $1 = 7 + b \rightarrow b = -6$

The linear equation is: $y = x - 6$

PRACTICES:

Write the equation of each line in form of $y = mx + b$.

1) $m = 2$; y-intercept$= -7$	2) $m = -\frac{2}{5}$; y-intercept$= \frac{2}{3}$
3) $f(-2) = 1; f(-3) = 4$	4) $f(6) = -1; f(2) = 7$
5) Through: $(5, 7), (3, 6)$	6) Through: $(-1.5, 2), (6.5, -2)$
7) Through: $(2, -1), (6, 11)$	8) Through: $(4, 1), (-2, 7)$
9) Through: $(2, 4), (-2, -4)$	10) Through: $(4, 5), (0, -1)$

Score: ....

Answer Key

1) $y = 2x - 7$	2) $y = -\frac{2}{5}x + \frac{2}{3}$
3) $y = -3x - 5$	4) $y = -2x + 11$
5) $y = \frac{1}{2}x + \frac{9}{2}$	6) $y = -0.5x + 1.25$
7) $y = 3x - 7$	8) $y = -x + 5$
9) $y = 2x$	10) $y = 1.5x - 1$

Name: ....

Slope-intercept form

- ✓ The slope-intercept form is one of several ways you can write the equation of a line.
- ✓ Using the slope *m* and the *y*-intercept *b*, then the equation of the line is:

$$y = mx + b$$

EXAMPLE:

Solve for y when 4x - 2y = 12.

Subtract $4x$ from both sides: $-4x + 4x - 2y = 12 - 4x \rightarrow -2y = 12 - 4x$

Divide everything by -2: $y = -6 + 2x \rightarrow y = 2x - 6$

PRACTICES:

Write the slope–intercept form of the equation of each line.

1) $y - 4 = x + 3$	2) $5x + 14 = -3y$
3) $18x - 12y = -6$	4) $7x - 4y + 25 = 0$
5) $-\frac{1}{3}y = -2x + 3$	6) $5 - y - 4x = 0$
7) $-y = -6x - 9$	8) $-2(7x + y) = 24$
9) $3(y + 3) = 2(x - 3)$	10) $\frac{3}{4}y + \frac{1}{4}x + \frac{5}{4} = 0$

Score: ....

Answer Key

1) $y = x + 7$	2) $y = -\frac{5}{3}x - \frac{14}{3}$
3) $y = \frac{3}{2}x + \frac{1}{2}$	4) $y = \frac{7}{4}x + \frac{25}{4}$
5) $y = 6x - 9$	6) $y = -4x + 5$
7) $y = 6x + 9$	8) $y = -7x - 12$
9) $y = \frac{2}{3}x - 5$	10) $y = -\frac{1}{3}x - \frac{5}{3}$

Name: ....

Point-slope form

✓ Using the slope m and a point (x_1, y_1) on the line, the equation of the line is:

$$(y - y_1) = m\,(x - x_1)$$

EXAMPLE:

Write the point-slope form of an equation of a line with a slope of 3 that passes through the point $(3, -2)$.

The slope is 3, so m = 3. We also know one point, so we know $x_1 = 3$ and $y_1 = -2$. Now we can substitute these values into the general point-slope equation.

$$y - y_1 = m(x - x_1) \rightarrow y - (-2) = \frac{3}{8}(x - 3) \rightarrow y + 2 = \frac{3}{8}(x - 3)$$

PRACTICES:

Write an equation in point–slope form for the line that passes through the given point with the slope provided.

1) $(2, -3), m = 4$	2) $(-7, 4), m = \frac{1}{5}$
3) $(0, -6), m = -2$	4) $(-a, b), m = m$
5) $(-9, 1), m = 3$	6) $(3, 0), m = -5$
7) $(-4, 11), m = \frac{1}{3}$	8) $(0, 11), (-2, 11)$
9) $\left(-\frac{1}{3}, 3\right), m = \frac{1}{5}$	10) $(0, 0), (1, -3)$

Score: ....

Answer Key	
1) $y + 3 = 4(x - 2)$	2) $y - 4 = \frac{1}{5}(x + 7)$
3) $y + 6 = -2x$	4) $y - b = m(x + a)$
5) $y - 1 = 3(x + 9)$	6) $y = -5(x - 3)$
7) $y - 11 = \frac{1}{3}(x + 4)$	8) $y - 11 = 0$
9) $y - 3 = \frac{1}{5}\left(x + \frac{1}{3}\right)$	10) $y = -3x$

Name: ....

Equation of Parallel or Perpendicular lines

✓ Parallel lines: The two lines will never intersect, and their slopes are identical. The only difference between the two lines is the y-intercept.

$\begin{cases} y = m_1 x + b_1 \\ y = m_2 x + b_2 \end{cases}$: Then, $m_1 = m_2$ and $b_1 \neq b_2$.

✓ Perpendicular Lines do intersect. Their intersection forms a right, or 90° angle. The slope of one line is the negative reciprocal of the slope of the other line.

$\begin{cases} y = m_1 x + b_1 \\ y = m_2 x + b_2 \end{cases}$: then, $m_1 = -\frac{1}{m_2}$, or $m_1 \times m_2 = -1$

EXAMPLE:

Given the functions below, identify the functions whose graphs are a pair of parallel lines

and a pair of perpendicular lines. $\begin{cases} f(x) = 4x - 5 \qquad\qquad g(x) = -4x + 2 \\ h(x) = \frac{1}{4}x - 5 \qquad\qquad p(x) = 4x + 7 \end{cases}$

Parallel lines have the same slope ($f(x)$, and $p(x)$)

Perpendicular lines have negative reciprocal slopes ($g(x)$, and $h(x)$).

PRACTICES:

Write an equation of the line that passes through the given point and is parallel to the given line.	Write an equation of the line that passes through the given point and is perpendicular to the given line.
1) $(0,7), -5x - y = -4$	6) $(\frac{3}{5}, \frac{2}{5}), y = -6x - 24$
2) $(-2,-1), y = \frac{4}{5}x + 3$	7) $(-10,0), y = \frac{5}{3}x - 15$
3) $(-2,5), -8x + 5y = -18$	8) $(3,-5), y = x + 12$
4) $(3,-2), y = -\frac{2}{5}x - 3$	9) $(-3,-1), y = \frac{7}{3}x - 4$
5) $(-5,-5), 6x + 15y = -30$	10) $(0,0), y - 8x + 6 = 0$

Score: ....

Answer Key	
1) $y = -5x + 7$	2) $y = \frac{4}{5}x + \frac{3}{5}$
3) $y = \frac{8}{5}x + \frac{41}{5}$	4) $y = -\frac{2}{5}x - \frac{4}{5}$
5) $y = -\frac{2}{5}x - 7$	6) $y = \frac{1}{6}x + \frac{3}{10}$
7) $y = -\frac{3}{5}x - 6$	8) $y = -x - 2$
9) $y = -\frac{3}{7}x - \frac{16}{7}$	10) $y = -\frac{1}{8}x$

Name: ....

Equation of Horizontal and Vertical Lines

✓ Equations of horizontal and vertical lines only have one variable.
✓ The slope of horizontal lines is 0 and y-values for each point are the same. Then, the equation of horizontal lines is: $y = b$.
✓ The slope of vertical lines is undefined and the equation for a vertical line is: $x = a$

EXAMPLE:

Write an equation for the vertical line that passes through $(4, -1)$.

As the line is vertical, x is constant, and x always takes the same value. Then x always takes the value 4. Thus, the equation is $x = 4$.

PRACTICES:

Sketch the graph of each line.

1) $y = 3$

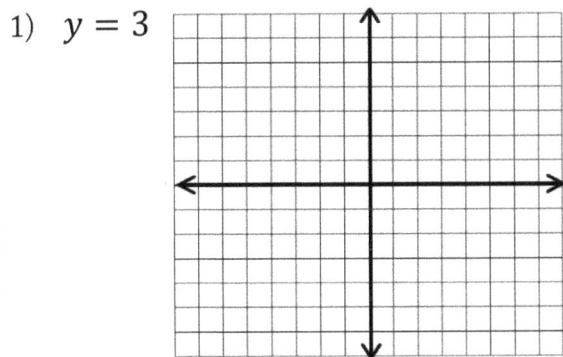

2) $y = -1$

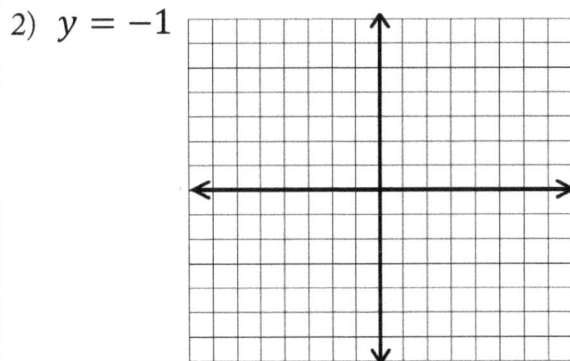

3) $x = 0$

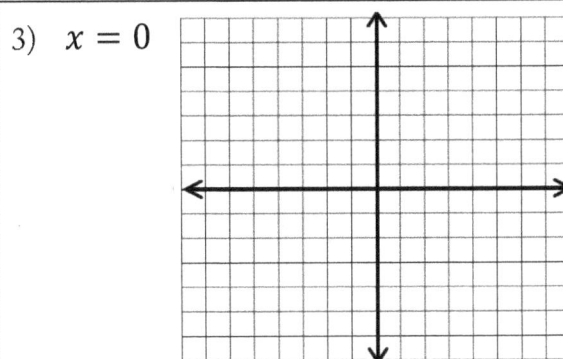

4) $x = 3$

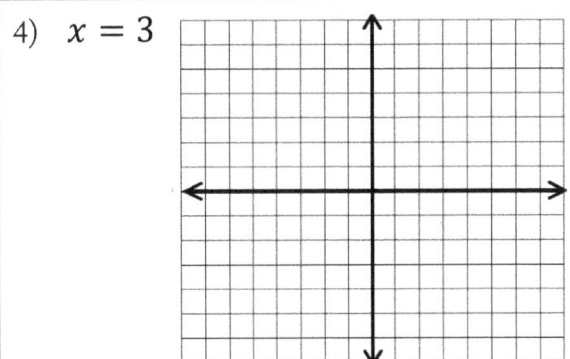

Score: ....

Answer Key

1) $y = 3$

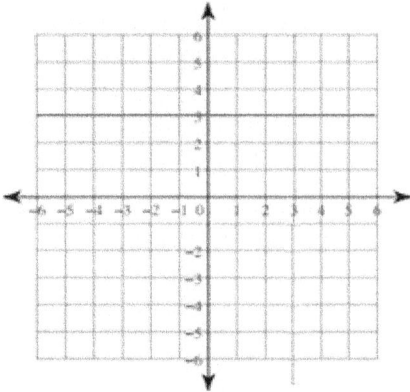

2) $y = -1$

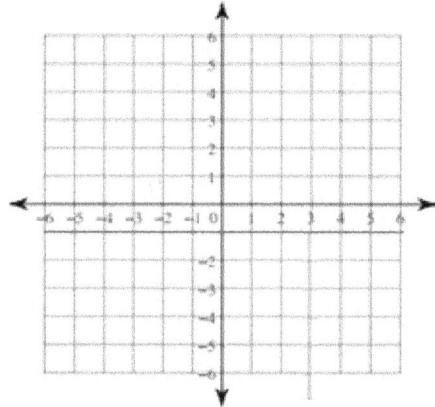

3) $x = 0$

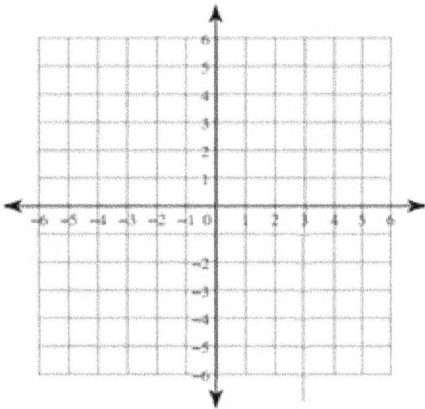

4) $x = 3$

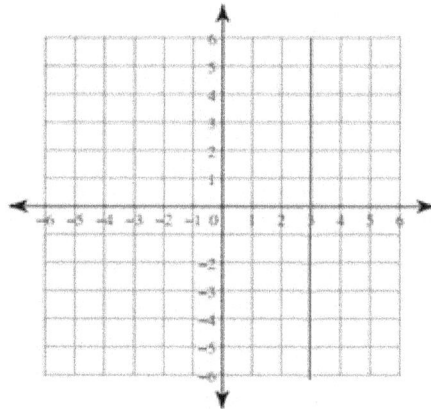

Name: ....

Function Notation

- ✓ **Function notation** is a way to write functions that is easy to read and understand. Also, it determines that a relationship is a function.
- ✓ The notation y = f (x) defines a function named f. This is read y is a function of x. The letter x is the input value or independent variable. The letter y, or f (x), is the output value or dependent variable.

EXAMPLE:

Write $y = x^2 + 3x - 5$ using function notation and evaluate the function at $x = 2$.

By applying function notation, we get: $f(x) = x^2 + 3x - 5$

Evaluation: Substitute x with 2:

$f(2) = 2^2 + 3 \times 2 - 5 = 4 + 6 - 5 = 5$

PRACTICES:

Write in function notation.

1) $v = 8t$	2) $r = 4p^2 + 2p - 2$
3) $h = 15g + 8$	4) $y = 5x - \frac{3}{4}$

Evaluate each function.

5) $h(x) = x^3 - 8$, find $h(-2)$	6) $f(u) = 9u - 2$, find $f(^1/_3)$
7) $h(x) = 3x - 6$, find $h(a)$	8) $h(a) = -2a + 4$, find $h(3b)$
9) $h(x) = x^2 + 4x - 7$, find $h(x^2)$	10) $h(x) = x^2 + 5$, find $h(-\frac{a}{3})$

Score: ....

Answer Key	
1) $v(t) = 8t$	2) $r(p) = 4p^2 + 2p - 2$
3) $h(g) = 15g + 8$	4) $f(x) = 5x - \frac{3}{4}$
5) -16	6) 1
7) $3a - 6$	8) $-6b + 4$
9) $x^4 + 4x^2 - 7$	10) $\frac{1}{9}a^2 + 5$

Name: ....

Adding and Subtracting Functions

✓ Like numbers and polynomials, we can add and subtract functions which results into a new function.

✓ Let f(x) and g(x) be two functions:

We can add two functions as: $(f + g)(x) = f(x) + g(x)$

We can subtract two functions as: $(f - g)(x) = f(x) - g(x)$

EXAMPLE:

Find the sum, and difference of $f = 3x + 1$ and $g = x - 2$ at the point 5.

$f + g = (3x + 1) + (x - 2) = 4x - 1 \rightarrow (f + g)(5) = 19$

$f - g = (3x + 1) - (x - 2) = 2x + 3 \rightarrow (f - g)(5) = 13$

PRACTICES:

Perform the indicated operation.

1) h(t) = 5t − 3 g(t) = 5t + 3 Find $(h - g)(t)$.	2) h(n) = 4n − 4 g(n) = n² − 6n + 9 Find (h + g)(a).
3) g(a) = −3a² + 4 f(a) = 2a² − a + 4 Find (g − f)(a).	4) g(x) = − x² + 8 − 3x f(x) = 8 + 2x Find (g − f)(x).
5) h(x) = 3x² − 5 g(x) = −4x² + 2x Find (h + g)(t).	6) g(t) = t + 8 f(t) = −3t² + t Find (g − f)(u − 1).
7) $h(x) = -3x + 4$ $g(x) = 2x - 6$ Find $(h + g)(2)$.	8) $k(x) = -3x + 6$ $h(x) = x^2 + 2x + 4$ Find $(k + h)(t - 3)$.

Score: ....

Answer Key	
1) -6	2) $a^2 - 2a + 5$
3) $-5a^2 + a$	4) $-x^2 - 5x$
5) $-t^2 + 2t - 5$	6) $3u^2 - 6u + 11$
7) -4	8) $t^2 - 7t + 22$

Name: ....

Multiplying and Dividing Functions

- ✓ Like numbers and polynomials, we can multiply and divide functions which results into a new function.
- ✓ Let f(x) and g(x) be two functions:

  We can multiply two functions as: $(f \cdot g)(x) = f(x) \cdot g(x)$

  We can divide two functions as: $\left(\frac{f}{g}\right)(x) = \frac{f(x)}{g(x)}$

EXAMPLE:

Given that $f(x) = x + 3$ and $g(x) = x^2 - 9$, find $(fg)(x)$ and $\left(\frac{f}{g}\right)(x)$ at the point 1.

$(fg)(x) = f(x) \times g(x) = (x+3)(x^2-9) = x^3 + 3x^2 - 9x - 27 \rightarrow (fg)(1) = -32$

$\left(\frac{f}{g}\right)(x) = \frac{f(x)}{g(x)} = \frac{x+3}{x^2-9} = \frac{x+3}{(x-3)(x+3)} = \frac{1}{x-3} \rightarrow \left(\frac{f}{g}\right)(1) = -\frac{1}{2}$

PRACTICES:

Perform the indicated operation.

1) $f(x) = x^2 - 3x$

   $g(x) = 4x^2 - 2$

   Find $(f.g)(x)$

2) $g(t) = \frac{1}{3}t^2 + \frac{1}{3}$

   $h(t) = 3t - 3$

   Find $(h.g)(\frac{1}{3})$

3) $f(a) = 12a - 8$

   $g(a) = 5a + 10$

   Find $(\frac{f}{g})(-1)$

4) $h(a) = -2a$

   $g(a) = -6a^2 - 2a$

   Find $(\frac{h}{g})(a)$

5) $g(a) = 4a - 3$

   $h(a) = 4a - 2$

   Find $(g.h)(2)$

6) $k(n) = 2n^2 - n$

   $h(n) = 3n^2 - 3$

   Find $(k.h)(1)$

7) $f(x) = 2x + 1$

   $g(x) = 4x^2 - 1$

   Find $(\frac{g}{f})(x)$

8) $f(t) = -a + 3$

   $g(t) = a^3 + 2$

   Find $(\frac{3f}{g})(a)$

Score: ....

Answer Key	
1) $4x^4 - 12x^3 - 2x^2 + 6x$	2) $-\dfrac{20}{27}$
3) -4	4) $\dfrac{1}{3a+1}$
5) 30	6) 0
7) $2x - 1$	8) $\dfrac{-3a+9}{a^3+2}$

Name: ....

Composition of Functions

- ✓ A composite function is generally a function that is written inside another function. Composition of a function is done by substituting one function into another function.

- ✓ The notation used for composition is:
$$(f o g)(x) = f(g(x))$$

EXAMPLE:

Using $f(x) = x + 1$ and $g(x) = 2x$, find: $(f\ o\ g)(1)$

$(f\ o\ g)(x) = f(g(x)) = 2(x + 1) = 2x + 2$
$(f\ o\ g)(1) = 4$

PRACTICES:

Using $f(x) = 2x - 8$, and $g(x) = -2x + 1$, find:

1) $f(g(0))$

2) $f(f(1))$

3) $g(f(3))$

Using $f(x) = 3x - 4a$, and $g(x) = x^2 - 2$, find:

4) $(fog)(1) = f(g(1))$

5) $(fof)(3)$

6) $(gof)(2)$

Using $f(x) = -2x + 3$, and $g(x) = x - b$, find:

7) $(fog)(-2x)$

8) $(fog)(x + 1)$

9) $(gof)(x^2)$

Score: ....

Answer Key	
1) -6	2) -20
3) 5	4) $-3 - 4a$
5) $27 - 16a$	6) $16a^2 - 48a + 34$
7) $4x + 3 + 2b$	8) $-2x + 1 + 2b$
9) $-2x^2 + 3 - b$	

Name: ....

Solve a Quadratic Equation

✓ Write the equation in the Standard form:

$ax^2 + bx + c = 0$ (One side must only contain zero)

✓ Factorize the quadratic.

✓ Use quadratic formula if you couldn't factorize the quadratic.

✓ Quadratic formula: $x = \dfrac{-b \pm \sqrt{b^2 - 4ac}}{2a}$

EXAMPLE:

Solve: $x^2 - 4x - 21 = 0$.

$$\begin{cases} a = 1 \\ b = -4 \\ c = -21 \end{cases} \Rightarrow x = \frac{-b \pm \sqrt{b^2 - 4ac}}{2a} = \frac{-(-4) \pm \sqrt{(-4)^2 - 4(1)(-21)}}{2(1)} = \begin{cases} \frac{4 + \sqrt{100}}{2} = 7 \\ \frac{4 - \sqrt{100}}{2} = -3 \end{cases}$$

This equation is also factorable:

$$x^2 - 4x - 21 = 0 \rightarrow (x - 7)(x + 3) = 0 \rightarrow \begin{cases} x - 7 = 0 \rightarrow x = 7 \\ x + 3 = 0 \rightarrow x = -3 \end{cases}$$

PRACTICES:

Solve each equation by using the quadratic formula.

1) $x^2 + 6x = -8$	2) $3x^2 - 9x - 9 = 3$
3) $6x^2 = 24x - 18$	4) $x^2 = 3x$
5) $3x^2 + 45 = -24x$	6) $2x^2 - 24x = -72$
7) $-6x^2 - 10x - 4 = 8 - 4x^2$	8) $x^2 - 20x = -84$
9) $2x^2 + 18x + 52 = 12$	10) $x^2 + 2x = 15 + 4x$

Score: ....

Answer Key	
1) $\{-4, -2\}$	2) $\{4, -1\}$
3) $\{1, 3\}$	4) $\{3, 0\}$
5) $\{-3, -5\}$	6) $\{6\}$
7) $\{-3, -2\}$	8) $\{14, 6\}$
9) $\{-4, -5\}$	10) $\{5, -3\}$

Chapter 9 : Geometry

Topics that you'll learn in this chapter:

➢ The Pythagorean Theorem

➢ Area of Triangles and Trapezoids

➢ Area and Circumference of Circles

➢ Area and Perimeter of Polygons

➢ Area of Squares, Rectangles, and Parallelograms

➢ Volume of Cubes, Rectangle Prisms, and Cylinder

➢ Surface Area of Cubes, Rectangle Prisms, and Cylinder

"Mathematics is, as it were, a sensuous logic, and relates to philosophy as do the arts, music, and plastic art to poetry." — *K. Shegel*

Name: ....

The Pythagorean Theorem

✓ In any right triangle: $a^2 + b^2 = c^2$

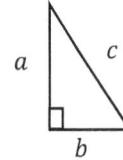

EXAMPLE:

Find the missing length.

Use Pythagorean Theorem: $a^2 + b^2 = c^2$

Then: $a^2 + b^2 = c^2 \rightarrow 3^2 + 4^2 = c^2 \rightarrow 9 + 16 = c^2$

$c^2 = 25 \rightarrow c = 5$

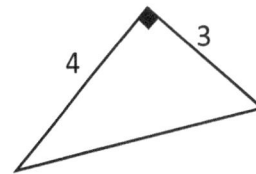

PRACTICES:

Do the following lengths form a right triangle?	Find each missing length to the nearest tenth.
1) 3, 5, 4	2) ?, 32, 60
3) 13, √290, 11	4) ?, 72, 35
5) 9, 12, 25	6) 55, 20, ?
7) 5, 13, 12	8) ?, 48, 64

9)

10)

Score: ....

Answer Key	
1) Yes	2) 68
3) Yes	4) 62.92
5) No	6) 58.52
7) Yes	8) 80
9) No	10) 12

Name: ....

Angles

✓ **Adjacent:** Two triangles are said to be adjacent if its two angles have common side, common vertex and do not overlap.

✓ **Vertical:** Two angles share same vertex.

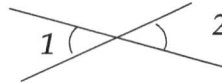

✓ **Complementary:** Sum of the measure of two complementary angles are 90°

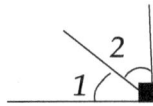

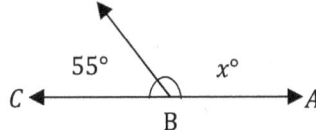

✓ **Supplementary:** Sum of the measure of two supplementary angles are 180°

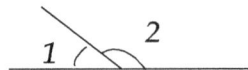

EXAMPLE:

Find x.

Supplementary: Sum of the measure of two supplementary angles are 180°.

Then: $180° - 55° = 125°$

PRACTICES:

What is the value of x in the following figures?

1)

2)

3)

4)

5)	6)

7)   8)

Solve.

9) Six supplement peer to each other angles have equal measures. What is the measure of each angle? _____

10) The measure of an angle is one fourth the measure of its complementary.What is the measure of the angle? _____

Score: ....

Answer Key	
1) 60°	2) 91°
3) 32°	4) 25°
5) 50°	6) 18°
7) 20°	8) 123°
9) 30°	10) 18°

Name: ....

Area of Triangles

- ✓ In any triangle the sum of all angles is 180 degrees.
- ✓ Area of a triangle = $\frac{1}{2}$ (base × height)

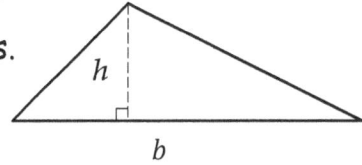

EXAMPLE:

What is the area of triangle?

Solution:

Use the are formula: Area = $\frac{1}{2}$ (base × height)

base = 12 and height = 8

Area = $\frac{1}{2}(12 \times 8) = \frac{1}{2}(96) = 48$

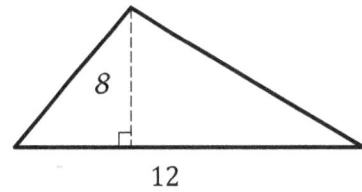

PRACTICES:

Find the area of each.

1)

c = 15 mi

h = 4 mi

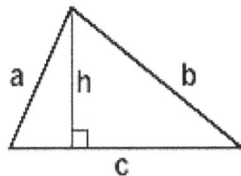

2)

c = 6 m

h = 5.2 m

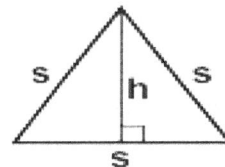

3)

a = 9.5 m

b = 25 m

c = 18 m

h = 9 m

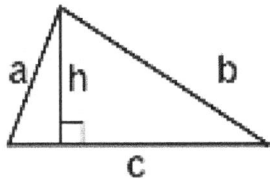

4)

s = 8 m

h = 6.93 m

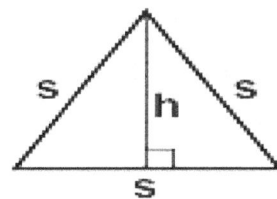

5)

c = 25 mi

h = 8 mi

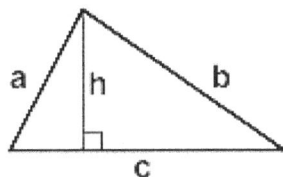

6)

c = 10 m

h = 6.4 m

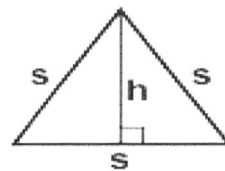

7)		8)	
a = 3.5 m b = 7 m c = 16 m h = 7 m		s = 12 m h = 4.64 m	

9)		10)	
c = 13 mi h = 6 mi		S = 18 m h = 7.4 m	

Score: ....

Answer Key

1) 30 mi²	2) 15.6 m²
3) 81 m²	4) 27.72 m²
5) 100 mi²	6) 32 m²
7) 56 m²	8) 27.84 m²
9) 39 mi²	10) 133.2 m²

Name: ....

Area of Trapezoids

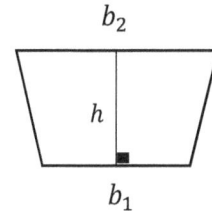

✓ A trapezoid is a quadrilateral with at least one pair of parallel sides.

✓ Area of a trapezoid = $\frac{1}{2}h(b_1 + b_2)$

EXAMPLE:

Calculate the area of the trapezoid.

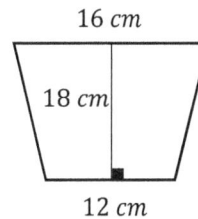

Use area formula: $A = \frac{1}{2}h(b_1 + b_2)$

$b_1 = 12$, $b_2 = 16$ and $h = 18$

Then: $A = \frac{1}{2}18(12 + 16) = 9(28) = 252\ cm^2$

PRACTICES:

Calculate the area for each trapezoid.

1)

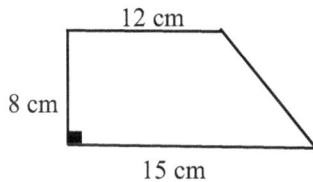

12 cm

8 cm

15 cm

2)

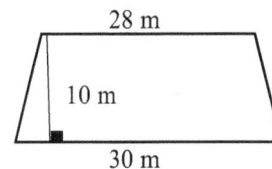

28 m

10 m

30 m

3)

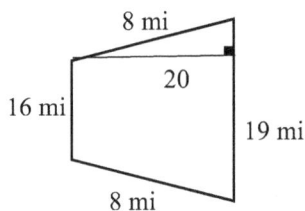

8 mi

20

16 mi

19 mi

8 mi

4)

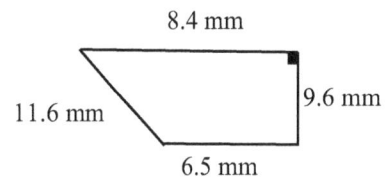

8.4 mm

11.6 mm

9.6 mm

6.5 mm

5)

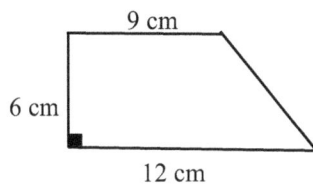

9 cm

6 cm

12 cm

6)

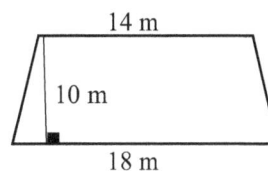

14 m

10 m

18 m

7)	8)
5 mi 7 mi 10 mi	7 mm 9.6 mm 4 mm 5 mm

9)	10)
13 cm 10 cm 18 cm	32 m 12 m 40 m

Score: ....

Answer Key

1) 108 cm²	2) 290 m²
3) 350 mi²	4) 71.52 mm²
5) 63 cm²	6) 160 m²
7) 42.5 mi²	8) 24 mm²
9) 155 cm²	10) 432 m²

Name: ....

Area and Perimeter of Polygons

Perimeter of a square
$= 4 \times side = 4s$

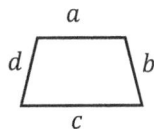
s

Perimeter of a rectangle
$= 2(width + length)$

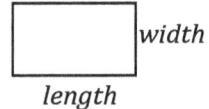
width

length

Perimeter of trapezoid
$= a + b + c + d$

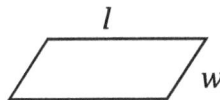
a
d
b
c

Perimeter of a regular hexagon $= 6a$

a

Perimeter of a parallelogram $= 2(l + w)$

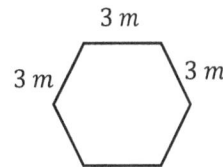
l
w

EXAMPLE:

Find the perimeter of following regular hexagon.

Perimeter of Pentagon $= 6a$

Perimeter of Pentagon $= 6a = 6 \times 3 = 18m$

3 m

3 m

3 m

PRACTICES:

Find the area and perimeter of each.

1)

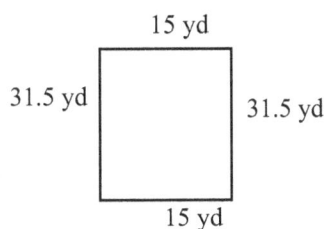
15 yd
31.5 yd
31.5 yd
15 yd

2)

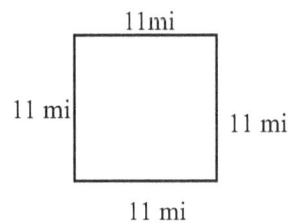
11mi
11 mi
11 mi
11 mi

3)

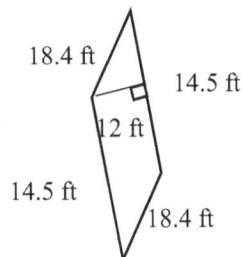
18.4 ft
14.5 ft
12 ft
14.5 ft
18.4 ft

4)

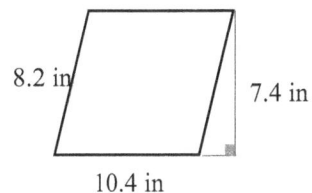
8.2 in
7.4 in
10.4 in

5)

15 cm ◄ ► 10 cm

13 cm

6)

5 mm

8 mm   6 mm

5 mm

Find the perimeter of each shape.

7)

6 m

6 m   6 m

8)

11 mm

11 mm

9)

13 ft   13 ft

10)

20 in

19 in

Score: ....

Answer Key

1) Area: $472.5\ yd^2$, Perimeter: 93 yd	2) Area: $121\ mi^2$, Perimeter: 44 mi
3) Area: $174\ ft^2$, Perimeter: 65.8 ft	4) Area: $76.96\ in^2$, Perimeter: 37.2 in
5) Area: $75cm^2$, Perimeter 52 cm	6) Area: $70\ mm^2$, Perimeter:38 mm
7) P: 36 m	8) P: 44 mm
9) P: 52 ft	10) P: 78 in

Name: ....

Area and Circumference of Circles

✓ We use variable r for the radius and d for diameter in a circle and π is about 3.14.
✓ Area of a circle$= \pi r^2$
✓ Circumference of a circle$= 2\pi r$

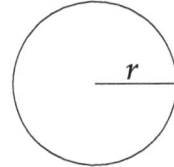

EXAMPLE:

Find the Circumference and area of the circle.

Use Circumference formula: $Circumference = 2\pi r$

$r = 6$, then: $Circumference = 2\pi(6) = 12\pi$

$\pi = 3.14$ then: $Circumference = 12 \times 3.14 = 37.68$

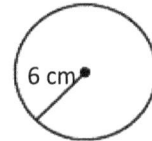

Use area formula: $Area = \pi r^2$,

$r = 6$ then: $Area = \pi(6)^2 = 36\pi$, $\pi = 3.14$ then: $Area = 36 \times 3.14 = 113.04$

PRACTICES:

Find the area and circumference of each. $(\pi = 3.14)$

1)

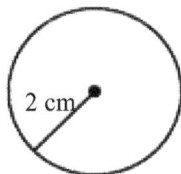

2)

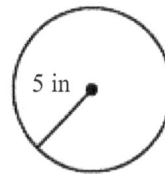

3)

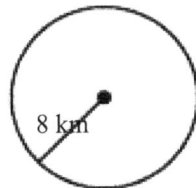

4)

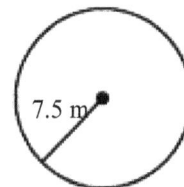

5) 4 m	6) 10 cm
7) 2.5cm	8) 1.5 in
9) 6 km	10) 24 in

Score: ....

Answer Key

1) Area: 12.56 cm², Circumference: 12.56cm.	2) Area: 78.5 in2, Circumference: 31.4 in.
3) Area: 200.96 km², Circumference: 50.24 km.	4) Area: 176.625 m2, Circumference: 47.1 m.
5) Area: 50.24 m², Circumference: 25.12 m	6) Area: 78.5 cm2, Circumference: 31.4 cm.
7) Area: 4.906 cm², Circumference: 7.85 cm.	8) Area: 1.766 in2, Circumference: 4.71 in.
9) Area: 113.04 km², Circumference:37.68 km.	10) Area: 452.16 in², Circumference: 75.36 in

Name: ....

Volume of Cubes

- ✓ A three-dimensional solid object bounded by six square sides is called cube.
- ✓ The measure of the amount of space inside of a solid figure is called volume, like a cube, ball, cylinder, or pyramid.
- ✓ Volume of a cube = $(one\ side)^3$

EXAMPLE:

Find the volume of this cube.

Use volume formula: $volume = (one\ side)^3$

Then: $volume = (one\ side)^3 = (2)^3 = 8\ cm^3$

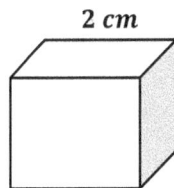

2 cm

PRACTICES:

Find the volume of each.

1)

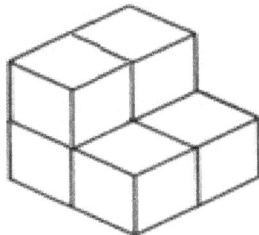

2)

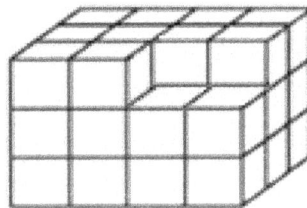

3)

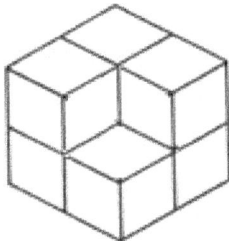

4)

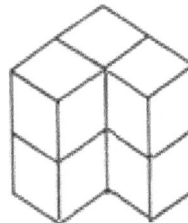

5)

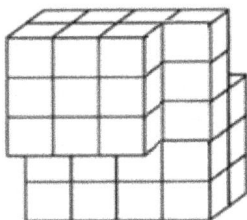

6)

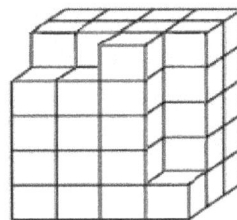

7) 4 ft	8) 6 m
9) 5 in	10) 3 miles

Score: ....

Answer Key	
1) 6	2) 34
3) 7	4) 6
5) 41	6) 54
7) $64\ ft^3$	8) $216\ m^3$
9) $125\ in^3$	10) $27\ mi^3$

Name: ....

Volume of Rectangle Prisms

✓ A solid 3-dimensional object which has six rectangular faces.

✓ Volume of a Rectangular prism = **Length × Width × Height**

Volume = $l \times w \times h$

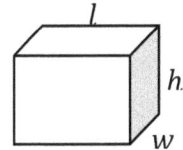

EXAMPLE:

Find the volume and surface area of rectangular prism.

Use volume formula: $Volume = l \times w \times h$

Then: $Volume = 10 \times 5 \times 8 = 400\ m^3$

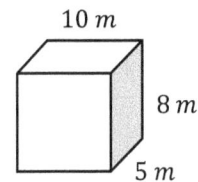

PRACTICES:

Find the volume of each of the rectangular prisms.

1)

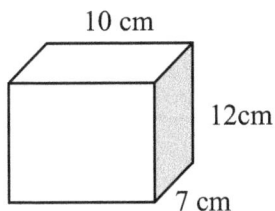

10 cm
12cm
7 cm

2)

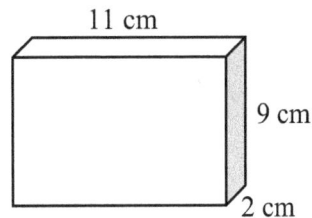

11 cm
9 cm
2 cm

3)

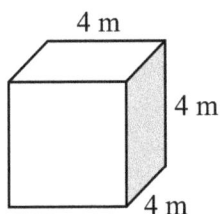

4 m
4 m
4 m

4)

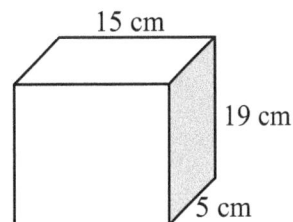

15 cm
19 cm
5 cm

5)

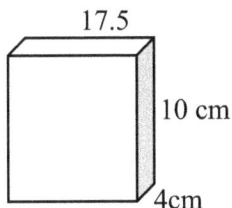

17.5
10 cm
4cm

6)

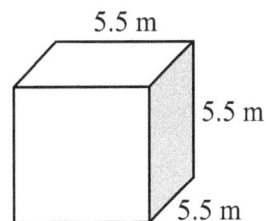

5.5 m
5.5 m
5.5 m

| 7) 7 m, 7 m, 7 m | 8) 20 ft, 11 ft, 3 ft |
| 9) 12.5 km, 8 km, 3 km | 10) 8 in, 10 in, 5 in |

Score: ....

Answer Key

1) 840 cm³	2) 198 cm³
3) 64 m³	4) 1,425 cm³
5) 700 cm³	6) 166.375 cm³
7) 343 cm³	8) 660 ft³
9) 300 km³	10) 400 in³

Name: ....

Surface Area of Cubes

✓ A three-dimensional solid object bounded by six square sides is called cube.

surface area of cube = $6 \times (one\ side)^2$

EXAMPLE:

Find the volume and surface area of this cube.

surface area of cube: $6(one\ side)^2 = 6(2)^2 = 6(4) = 24\ cm^2$

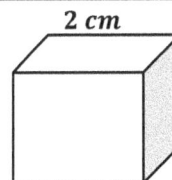

2 cm

PRACTICES:

Find the surface of each cube.

1) 7 mm

2) 10.5 mm

3) 3.5 cm

4) 4 m

5) 3.2 in

6) 8.1 ft

7) 1.6 in	8) 11 m
9) 5.2 in	10) 2.25 mm

Score: ....

Answer Key

1) 294 mm²	2) 661.5 mm²
3) 73.5 cm²	4) 96 m²
5) 61.44 in²	6) 393.66 ft²
7) 15.36 in²	8) 726 m²
9) 162.24 in²	10) 30.375 mm²

Name: ....

Surface Area of a Rectangle Prism

✓ A solid 3-dimensional object which has six rectangular faces.
  Surface area= $2(wh + lw + lh)$

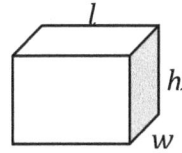

EXAMPLE:

Find the volume and surface area of rectangular prism.

Use surface area formula: $Surface\ area = 2(wh + lw + lh)$

Then: $Surface\ area = 2(5 \times 8 + 10 \times 5 + 10 \times 8) =$

$2(40 + 50 + 80) = 340\ m^2$

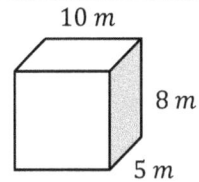

PRACTICES:

Find the surface of each prism.

1)
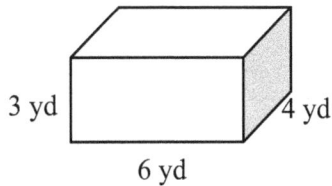
3 yd   4 yd
6 yd

2)
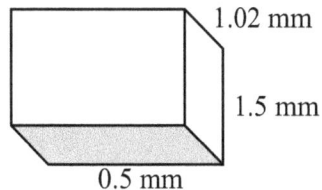
1.02 mm
1.5 mm
0.5 mm

3)
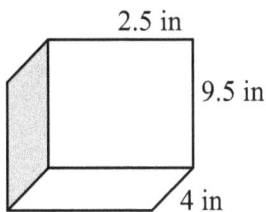
2.5 in
9.5 in
4 in

4)
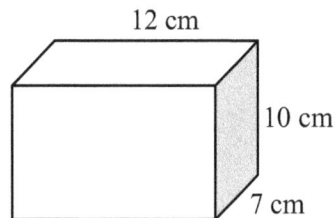
12 cm
10 cm
7 cm

5)
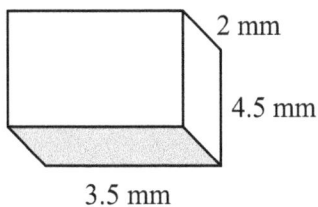
2 mm
4.5 mm
3.5 mm

6)
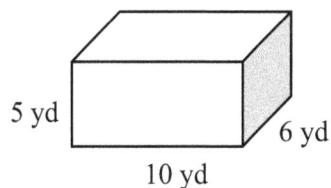
5 yd   6 yd
10 yd

7)

14 cm
8 cm
5 cm

8)

2 yd
6 yd
5 yd

9)

3.5 in
7.5 in
6 in

10)

4 mm
8 mm
6 mm

Score: ....

Answer Key

1) 108 yd²	2) 5.58 mm²
3) 143.5 in²	4) 548 cm²
5) 63.5 mm²	6) 280 yd²
7) 444 cm²	8) 104 yd²
9) 184.5 in²	10) 208 mm²

Name: ....

Volume of a Cylinder

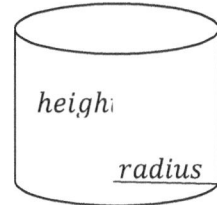

- ✓ A solid geometric figure with straight parallel sides and a circular cross section is called a cylinder.
- ✓ Volume of Cylinder Formula = $\pi(radius)^2 \times height$
  $\pi = 3.14$

EXAMPLE:

Find the volume of the follow Cylinder.

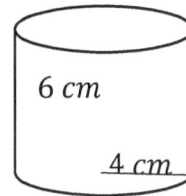

Use volume formula: $Volume = \pi(radius)^2 \times height$

Then: $Volume = \pi(4)^2 \times 6 = \pi16 \times 6 = 96\pi$

$\pi = 3.14$ then: $Volume = 96\pi = 301.44$

PRACTICES:

Find the volume of each cylinder. ($\pi = 3.14$).

1)

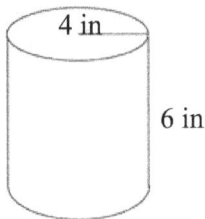

4 in
6 in

2)

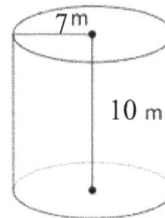

7 m
10 m

3)

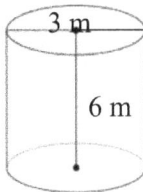

3 m
6 m

4)

2 in
4.5 in

5)

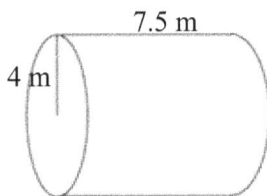

7.5 m
4 m

6)

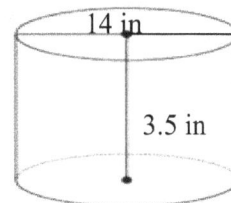

14 in
3.5 in

7) 10 in · 7.5 in	8) 6 ft · 10 ft
9) 5 in · 9 in	10) 12 yd · 2 yd

Score: ....

Answer Key	
1) 301.44 in³	2) 1538.6 m³
3) 42.39 m³	4) 14.13 in³
5) 376.8 m³	6) 538.51 in³
7) 588.75 in³	8) 282.6 ft³
9) 706.5 in³	10) 150.72 yd³

Name: ....

Surface Area of a Cylinder

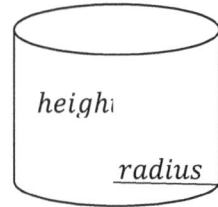

- ✓ A solid geometric figure with straight parallel sides and a circular cross section is called cylinder.
- ✓ Surface area of a cylinder= $2\pi r^2 + 2\pi rh$

height

radius

EXAMPLE:

Find the Surface area of the follow Cylinder.

Use surface area formula: $Surface\ area = 2\pi r^2 + 2\pi rh$

Then: $= 2\pi(4)^2 + 2\pi(4)(6) = 2\pi(16) + 2\pi(24) = 32\pi + 48\pi = 80\pi$

$\pi = 3.14$ then: $Surface\ area = 80 \times 3.14 = 251.2$

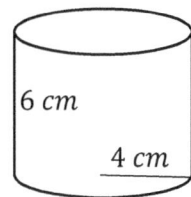

6 cm

4 cm

PRACTICES:

Find the surface of each cylinder. ($\pi = 3.14$).

1)

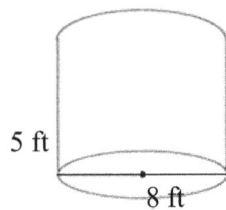

5 ft

8 ft

2)

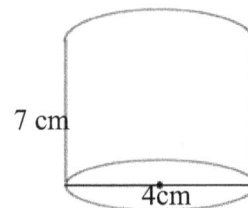

7 cm

4 cm

3)

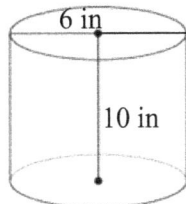

6 in

10 in

4)

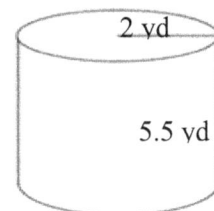

2 yd

5.5 yd

5)

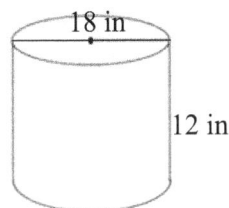

18 in

12 in

6)

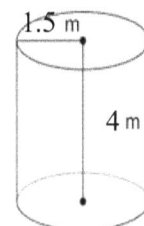

1.5 m

4 m

| 7) 6 in / 10 in | 8) 8 ft / 4 ft |
| 9) 1.4 yd / 6.6 yd | 10) 10 in / 20 in |

Score: ....

Answer Key	
1) 226.08 ft²	2) 113.04 cm²
3) 224.92 in²	4) 94.2 yd²
5) 1,186.92 in²	6) 51.81 m²
7) 244.92 in²	8) 502.4 ft²
9) 70.336 yd²	10) 785 in²

Chapter 10 : Statistics

Topics that you'll learn in this chapter:

> ➢ Mean, Median, Mode, and Range of the Given Data

> ➢ Box and Whisker Plots

> ➢ Bar Graph

> ➢ Stem– And– Leaf Plot

> ➢ The Pie Graph or Circle Graph

> ➢ Dot and Scatter Plots

> ➢ Probability of Simple Events

"The book of nature is written in the language of Mathematic" -Galileo

Name: ....

Mean and Median

✓ Mean: $\dfrac{\text{sum of the data}}{\text{total number of data entires}}$

✓ Median: Middle value in the sorted list of numbers

✓ When there are two middle numbers, we average them

EXAMPLE:

What is the median of these numbers? **4, 9, 13, 8, 15, 18, 5, 11**

Write the numbers in order: 4, 5, 8, 9, 11, 13, 15, 18

Median is the number in the middle. Therefore, there are 9 *and* 11 in the middle, then

find the average: $\dfrac{9+11}{2} = \dfrac{20}{2} = 10$, the median is 10

PRACTICES:

Find Mean and Median of the Given Data.

1) 8, 10, 7, 3, 12	2) 4, 6, 9, 7, 5, 19
3) 5, 11, 1, 1, 8, 9, 20	4) 12, 4, 2, 7, 3, 2
5) 3, 5, 7, 4, 7, 8, 9	6) 5, 10, 4, 4, 9, 12, 9
7) 10, 4, 8, 5, 9, 6, 7, 19	8) 16, 3, 4, 3, 7, 6, 18

Solve.

9) In a javelin throw competition, five athletics score 23, 45, 53. 53, 13 and 61 meters. What are their Mean and Median? _____

10) Eva went to shop and bought 7 apples, 4 peaches, 6 bananas, 3 pineapples and 4 melons. What are the Mean and Median of her purchase? _____

Score: ....

Answer Key

1) Mean: 8, Median: 8	2) Mean: 8.33, Median: 6.5
3) Mean: 7.85, Median: 8	4) Mean: 5, Median: 3.5
5) Mean: 6.14, Median: 7	6) Mean: 7.57, Median: 9
7) Mean: 8.5, Median: 7.5	8) Mean: 8.14, Median: 6
9) Mean: 39.106, Median: 45	10) Mean: 4.8, Median: 4

Name: ....

Mode and Range

✓ Mode: The most appeared value in the list.

✓ Range: The difference of highest value and lowest value in the list

EXAMPLE:

What is the mode(s) of these numbers? **22, 16, 12, 9, 7, 6, 4, 6, 9**

Mode: The most appeared value in the list.

Therefore: modes are 6 and 9

PRACTICES:

Find Mode and Rage of the Given Data.

1) 10, 12, 8, 8,4, 1, 9 Mode: _____ Range: _____	2) 4, 6, 4, 13, 2, 13, 19, 13 Mode: _____ Range: _____
3) 8, 8, 7, 2, 7, 7, 5, 6, 5 Mode: _____ Range: _____	4) 12, 9, 12,6, 12, 9, 10 Mode: _____ Range: _____
5) 2, 2, 4, 3, 2, 10, 8 Mode: _____ Range: _____	6) 6, 1, 4, 20, 19, 2, 7, 1, 5, 1 Mode: _____ Range: _____
7) 16, 35, 9, 7, 7, 5, 14, 13, 7 Mode: _____ Range: _____	8) 7, 6, 6, 9, 16, 6, 7, 5 Mode: _____ Range: _____

Solve.

9) A stationery sold 15 pencils, 26 red pens, 22 blue pens, 10 notebooks, 12 erasers, 22 rulers and 42 color pencils. What are the Mode and Range for the stationery sells?

Mode: _____ Range: _____

10) In an English test, eight students score 24, 13, 17, 21, 19, 13, 13 and 17. What are their Mode and Range? _____

Score: ....

Answer Key	
1) Mode: 8, Range: 11	2) Mode: 13, Range: 17
3) Mode: 7, Range: 6	4) Mode: 12, Range: 6
5) Mode: 2, Range: 8	6) Mode: 1, Range: 19
7) Mode: 7, Range: 30	8) Mode: 6, Range: 11
9) Mode: 22, Range: 32	10) Mode: 13, Range: 11

Name: ....

Times Series

✓ A precise representation of the distribution of numerical data is referred as Time Series.

EXAMPLE:

Use the following Graph to complete the table.

Day	Distance (km)
1	
2	

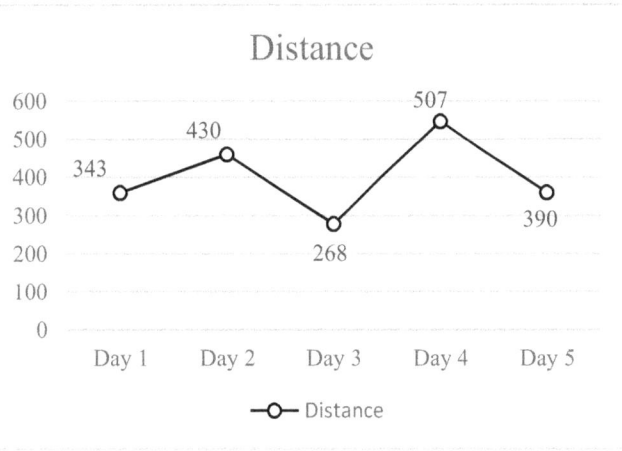

Answer:

Day	Distance (km)
1	359
2	460
3	278
4	547
5	360

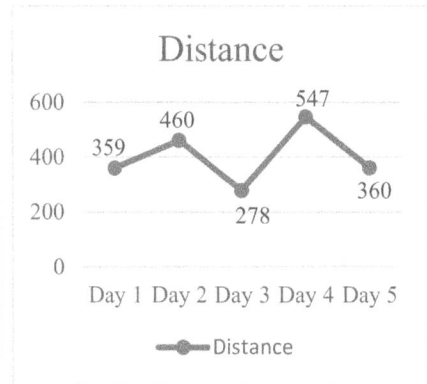

→

Distance

600
460   547
400   359
200   278   360
0
Day 1 Day 2 Day 3 Day 4 Day 5
━●━Distance

PRACTICES:

Use the following Graph to complete the table.

1)

Day	Distance (km)
1	
2	

Distance

600
500   430   507
400   343
300   268   390
200
100
0
Day 1   Day 2   Day 3   Day 4   Day 5
━○━ Distance

The following table shows the number of births in the US from 2007 to 2012 (in millions).

2)

Draw a time series for the table.

Year	Number of births (in millions)
2007	6.42
2008	6.45
2009	6.33
2010	5.9
2011	4.35
2012	4.35

Score: ....

Answer Key

1)

Day	Distance (km)
1	343
2	430
3	268
4	507
5	390

2)

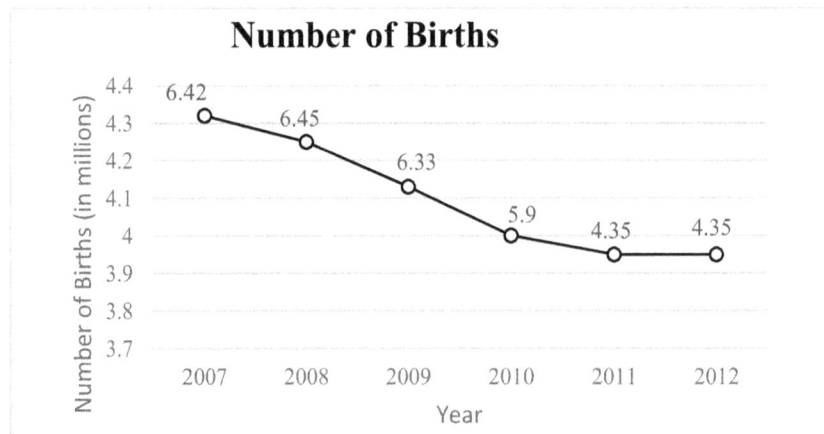

Number of Births

Name: ....

Box and Whisker Plot

✓ Box-and-whisker plots display data including quartiles.

✓ IQR – interquartile range shows the difference from Q1 to Q3.

✓ Extreme Values are the lowest and highest values in a data set.

EXAMPLE:

$73, 84, 86, 95, 68, 67, 100, 94, 77, 80, 62, 79$

Maximum: 100, Minimum: 62; Q_1: 70.5; Q_2: 79.5; Q_3: 90

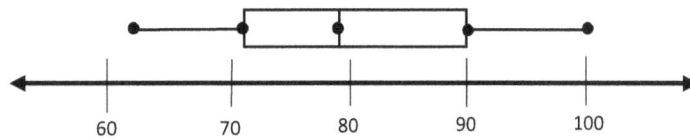

PRACTICES:

Make box and whisker plots for the given data.

1) $1, 5, 20, 8, 3, 10, 13, 11, 14, 17, 18, 15, 23$

2) $2, 7, 23, 11, 13, 9, 16, 5, 18, 22, 20, 17, 19$

3) $3, 7, 9, 10, 11, 5, 14, 19, 20, 21, 22, 8, 14$

4) $4, 6, 5, 15, 12, 14, 10, 7, 21, 17, 8, 22, 6$

Score: ....

Answer Key

1) 1,3, 5, 8, 10, 11, 13, 14, 15, 17,18, 20, 23

   Maximum: 23, Minimum: 1, Q_1: 8, Q_2: 13, Q_3: 17

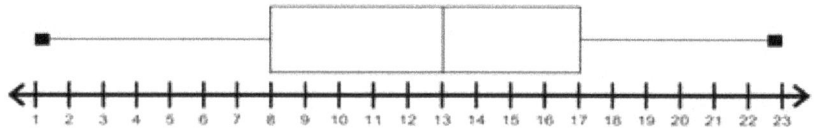

2) 2, 7, 23, 11, 13, 9, 16, 5, 18, 22, 20, 17, 19

   Maximum: 23, Minimum: 2, Q_1: 9, Q_2: 16, Q_3: 19

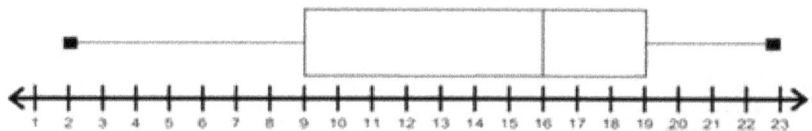

3) 3,7, 9, 10, 11, 5, 14, 19, 20, 21 ,22, 8, 14

   Maximum: 22, Minimum: 2, Q_1: 8, Q_2: 11, Q_3: 19

4) 4, 6, 5, 15, 12, 14, 10, 7, 21, 17, 8, 22, 6

   Maximum: 22, Minimum: 4, Q_1: 6, Q_2: 10, Q_3: 15

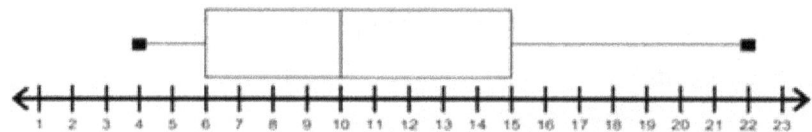

Name: ....

Bar Graph

✓ A chart that presents data with bars in different heights to match with the values of the data is called a bar graph. We can graph the bars vertically or horizontally.

EXAMPLE:

Graph the given information as a bar graph.

Name of the Sport	Total Number of Students
Football	15
Volleyball	7
Table Tennis	7
Basketball	12
Chess	9

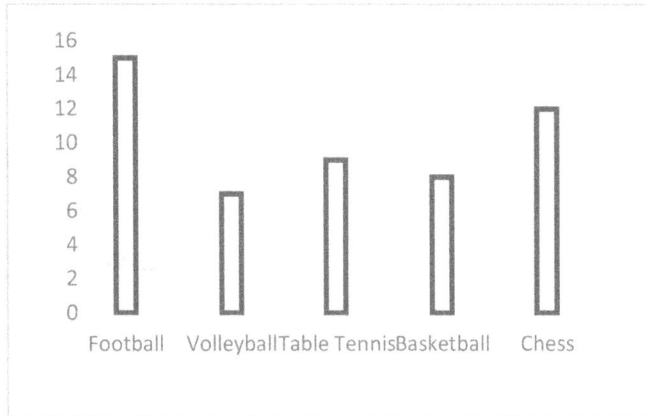

PRACTICES:

Graph the given information as a bar graph.

1)

Day	Sale House
Monday	6
Tuesday	4
Wednesday	10
Thursday	5
Friday	2
Saturday	8
Sunday	1

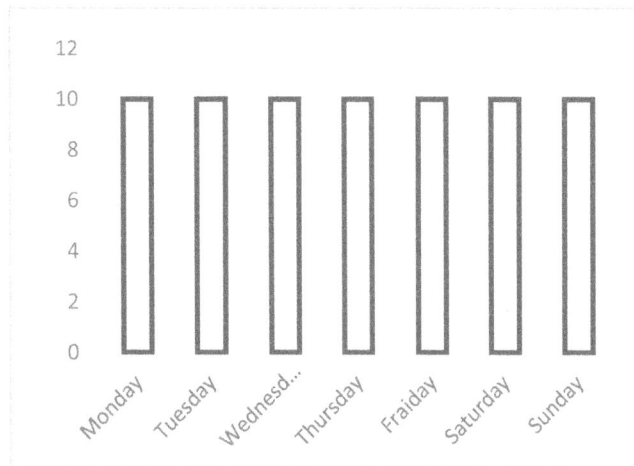

2)

Day	Sale House
Monday	8
Tuesday	6
Thursday	3
Friday	10
Saturday	4
Sunday	2

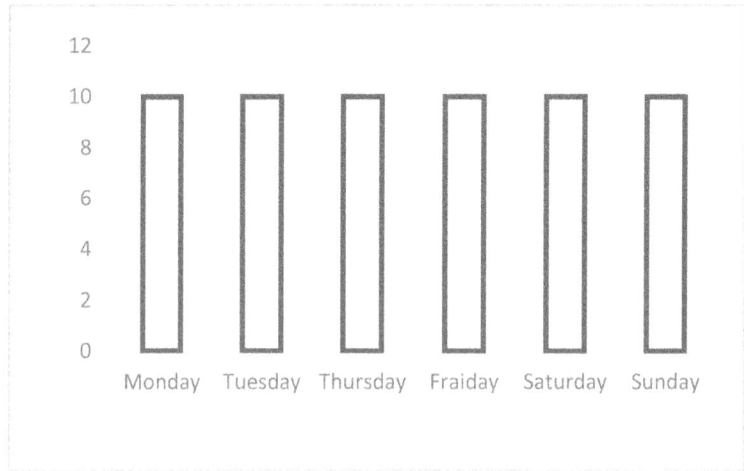

Score: ....

Answer Key

1)

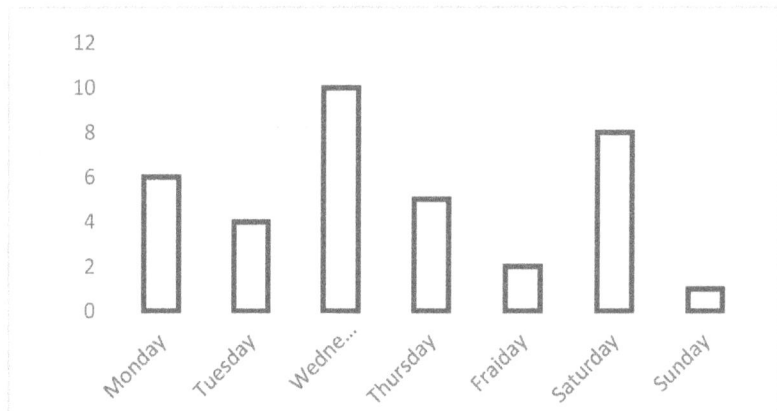

2)

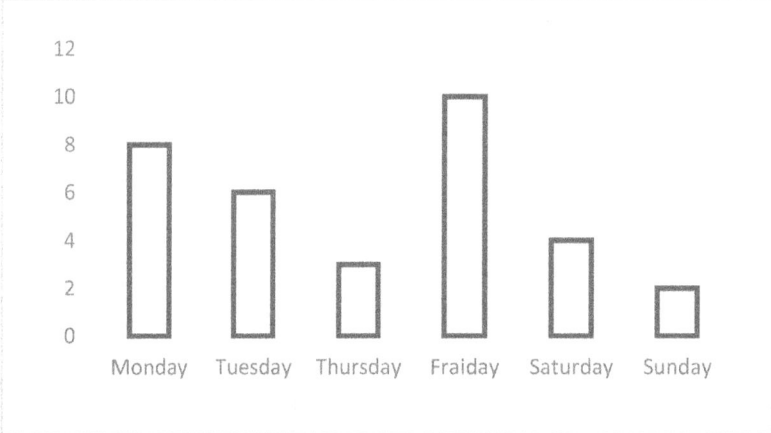

Name: ....

Dot plots

- ✓ The representation of a distribution that consists of group of data points plotted on a simple scale is referred as a dot plot. Dot plots are used for continuous, univariate and quantitative data. If there are few data points, then they can be labelled.
- ✓ Dot plots are one of the simplest statistical plots and are suitable for small to moderate sized data sets.

EXAMPLE:

A survey of "How many books each student purchased?" has these results How many students purchase 4 books?

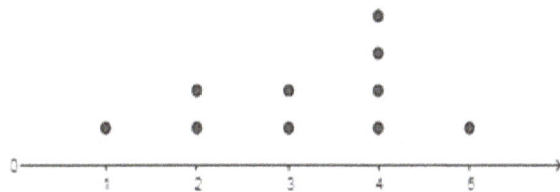

4 students

PRACTICES:

A survey of "How many pets each person owned?" has these results:

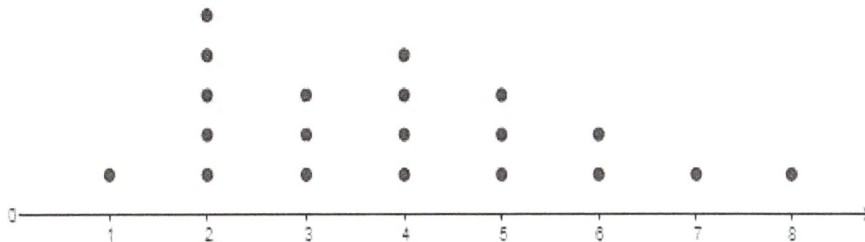

1) How many people have at least 3 pets?

2) How many people have 2 and 3 pets?

3) How many people have 4 pets?

4) How many people have 2 or less than 2 pets?

5) How many people have more than 7 pets?

6) How many people have more than 4 pets?

Score: ....

Answer Key	
1) 4	2) 8
3) 4	4) 6
5) 1	6) 7

Name: ....

Scatter Plots

- ✓ The values with points that represent the relationship between two sets of data are shown by a scatter(x, y) plot.
- ✓ The horizontal values are taken as x and vertical data is taken as y.

EXAMPLE:

Construct a scatter plot.

x	1	2	3	4	5
y	2	4.5	1.5	5	2

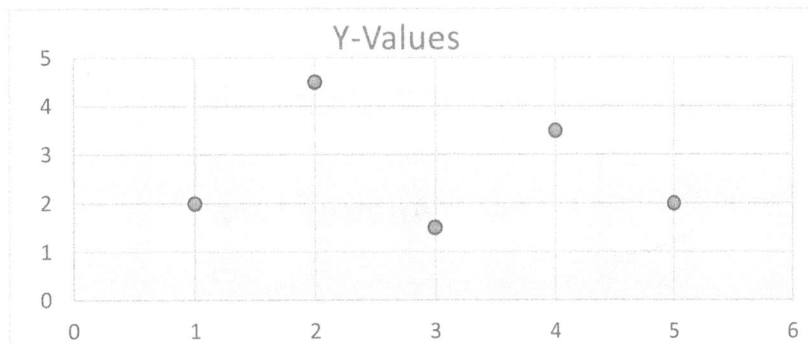

PRACTICES:

Construct a scatter plot.

1)

x	1	2.5	3	3.5	4	5
y	4	3.5	4.5	2.5	8	2

2)

x	1	2	3.5	4	4.5	5
y	3	1	2.5	1.5	1.5	1

Y-Values

Score: ....

Answer Key

1)

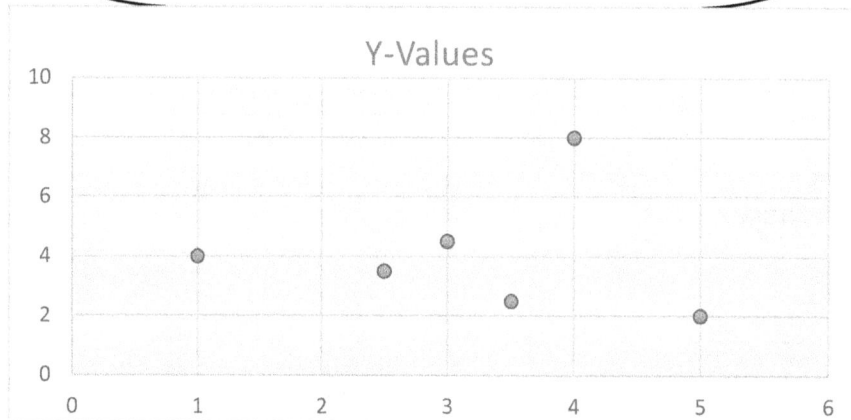

Y-Values

2)

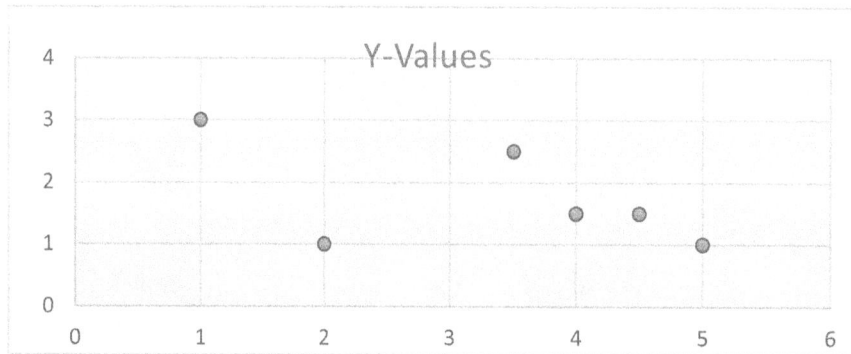

Y-Values

Name: ....

Stem–And–Leaf Plot

- ✓ Stem-and-leaf plots display the frequency of the values in a data set.
- ✓ We can make a frequency distribution table for the values, or we can use a stem-and-leaf plot.

EXAMPLE:

56, 58, 42, 48, 66, 64, 53, 69, 45, 72

Stem	leaf
4	2 5 8
5	3 6 8
6	4 6 9
7	2

PRACTICES:

Make stem ad leaf plots for the given data.

1) 42, 14, 17, 21, 44, 24, 18, 47, 23, 24, 19, 12

Stem	Leaf plot

2) 10, 65, 14, 18, 69, 11, 33, 61, 66, 38, 15, 35

Stem	Leaf plot

3) 122, 87, 99, 86, 100, 126, 92, 129, 88, 121, 91, 107

Stem	Leaf plot

4) 60, 51, 119, 69, 72, 59, 110, 65, 77, 59, 65, 112, 71

Stem	Leaf plot

Score: ....

Answer Key

1)

Stem	leaf
1	2 4 7 8 9
2	1 3 4
4	2 4 7

2)

Stem	leaf
1	0 1 4 5 8
3	3 5 8
6	1 5 6 9

3)

Stem	leaf
8	6 7 8
9	1 2 9
10	0 7
12	1 2 6 9

4)

Stem	leaf
5	1 9 9
6	0 5 5 9
7	1 2 7
11	0 2 9

Name: ....

The Pie Graph or Circle Graph

✓ A circular chart divided into sectors is said to be a pie chart; each sector represents the relative size of each value.

EXAMPLE:

A library has 840 books that include Mathematics, Physics, Chemistry, English, and History. Use following graph to answer question.

What is the number of Mathematics books?

Number of total books = 840,

Percent of Mathematics books = 30% = 0.30

Then: $0.30 \times 840 = 252$

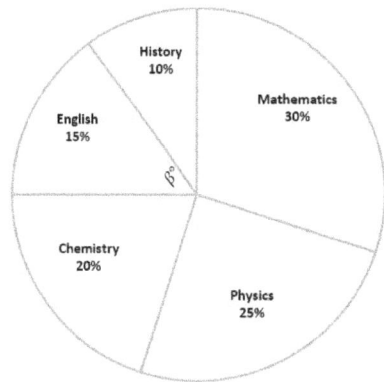

PRACTICES:

Favorite Sports:

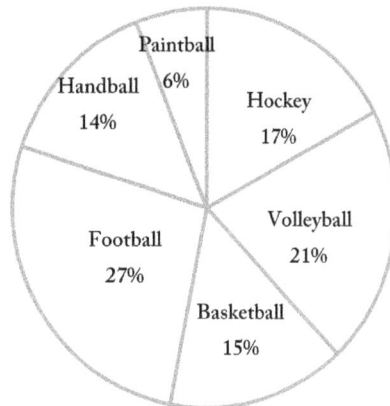

SPORTS

1) What percentage of pie graph is paintball?
2) What percentage of pie graph is Hockey and volleyball?
3) What percentage of pie not Football and Basketball?
4) What percentage of pie graph is Hockey and Handball and Football?
5) What percentage of pie graph is Basketball?
6) What percentage of pie not Handball and Paintball?

Score: ....

Answer Key	
1) 6%	2) 38%
3) 58%	4) 58%
5) 15%	6) 80%

Name: ....

Probability of Simple Events

- ✓ Probability is the possibility of something happening in the future. It is shown as a number between zero (can never happen) to 1 (will always happen).
- ✓ Probability can be written as a fraction, a decimal, or a percent.

EXAMPLE:

If there are 8 red balls and 12 blue balls in a basket, what is the probability that John will pick out a red ball from the basket?

There are 8 red ball and 20 are total number of balls. Therefore, probability that John will pick out a red ball from the basket is 8 out of 20 or $\frac{8}{8+12} = \frac{8}{20} = \frac{2}{5}$

PRACTICES:

Solve.

1) A number is chosen at random from 28 to 35. Find the probability of selecting factors of 5.

2) A number is chosen at random from 1 to 60. Find the probability of selecting multiples of 15.

3) Find the probability of selecting 4queens from a deck of card.

4) A number is chosen at random from 8 to 19. Find the probability of selecting factors of 3.

5) What probability of selecting a ball less than 6 from 10 different bingo balls?

6) A number is chosen at random from 1 to 10. What is the probability of selecting a multiple of 2?

7) A card is chosen from a well-shuffled deck of 52 cards. What is the probability that the card will be a king OR a queen?

8) A number is chosen at random from 1 to 20. What is the probability of selecting multiples of 5?

9) A number is chosen at random from 1 to 10. Find the probability of selecting number 4 or smaller numbers.

10) Bag A contains 9 red marbles and 3 green marbles. Bag B contains 9 black marbles and 6 orange marbles. What is the probability of selecting a green marble at random from bag A? What is the probability of selecting a black marble at random from Bag B?

Score: ....

Answer Key

1) $\frac{1}{4}$	2) $\frac{1}{15}$
3) $\frac{1}{13}$	4) $\frac{1}{3}$
5) $\frac{1}{2}$	6) $\frac{1}{2}$
7) $\frac{2}{13}$	8) $\frac{1}{5}$
9) $\frac{2}{5}$	10) $\frac{1}{4}, \frac{3}{5}$

Chapter 11 : PARCC Math Practice Tests

Time to Test

Time to refine your skill with a practice examination

Take a REAL PARCC Mathematics test to simulate the test day experience.

After you've finished, score your test using the answer key.

Before You Start

- You'll need a pencil, calculator, and a timer to take the test.

- It's okay to guess. You won't lose any points if you're wrong.

- After you've finished the test, review the answer key to see where you went wrong.

Graphing calculators are permitted for PARCC Tests Grade 8

Good Luck!

PARCC Grade 8 Mathematics Reference Materials

Linear Equations

Slope-intercept form	$y = mx + b$
Slope of a line	$m = \frac{y_2 - y_1}{x_2 - x_1}$

Circumference

Circle	$C = 2\pi r$	or	$C = \pi d$

Area

Triangle	$A = \frac{1}{2}bh$
Rectangle or Parallelogram	$A = bh$
Trapezoid	$A = \frac{1}{2}h(b_1 + b_2)$
Circle	$A = \pi r^2$

Surface Area

	Lateral	Total
Prism	$S = ph$	$S = ph + 2B$
Cylinder	$S = 2\pi rh$	$S = 2\pi rh + 2\pi r^2$

Volume

Prism or cylinder	$V = Bh$
Pyramid or Cone	$V = \frac{1}{3}Bh$
Sphere	$V = \frac{4}{3}\pi r^3$

Additional Information

Pythagorean theorem	$a^2 + b^2 = c^2$
Simple interest	$I = prt$
Compound interest	$I = p(1 + r)^t$

Partnership for Assessment of Readiness for College and Careers

PARCC Practice Test 1

Mathematics

GRADE 8

Unit 1

Calculators are NOT permitted for unit 1 of the test.

Read each question. Then mark your answers in your answer sheet.

If you have time, review your answers, and only answer questions you did not answer in the unit.

Time for Unit 1: 80 Minutes

Administered Month Year

1) Right triangle ABC has two legs of lengths 33 cm (AB) and 44 cm (AC). What is the length of the third side (BC)?

    A. 55 cm   C. 77 cm

    B. 54 cm   D. 40 cm

2) When a number is subtracted from 90 and the difference is divided by that number, the result is 8. What is the value of the number?

    A. 12   C. 8

    B. 10   D. 14

3) A bank is offering 3.75% simple interest on a savings account. If you deposit $8,000, how much interest will you earn in six years?

    A. $1,540   C. $1,800

    B. $1,350   D. $1,350

4) In a party, 30 soft drinks are required for every 15 guests. If there are 158 guests, how many soft drinks is required?

    A. 150   C. 316

    B. 335   D. 225

5) A chemical solution contains 12% alcohol. If there is 34.8 ml of alcohol, what is the volume of the solution?

    A. 380 ml   C. 390 ml

    B. 290 ml   D. 280 ml

6) What is the area of the shaded region?

   A. 107

   B. 112

   C. 35

   D. 77

(Diagram: outer rectangle 14 ft by 8 ft; inner rectangle 7 ft by 5 ft)

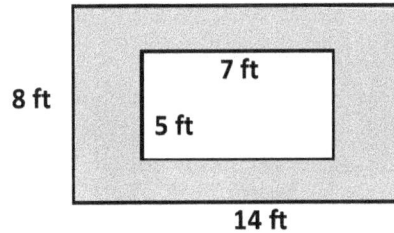

7) A rope weighs 300 grams per meter of length. What is the weight in kilograms of 46.4 meters of this rope? (1 kilograms = 1000 grams)

   A. 20.63　　　　　　　　　　C. 13.82

   B. 23.92　　　　　　　　　　D. 60.30

8) The ratio of boys to girls in a school is 5:7. If there are 624 students in a school, how many boys are in the school.

Write your answer in the box below.

[   ]

9) In two successive years, the population of a town is increased by 10% and 25%. What percent of the population is increased after two years?

   A. 28.6%　　　　　　　　　　C. 137.5%

   B. 37.5%　　　　　　　　　　D. 35.7%

10) Which graph shows a non–proportional linear relationship between x and y?

A.

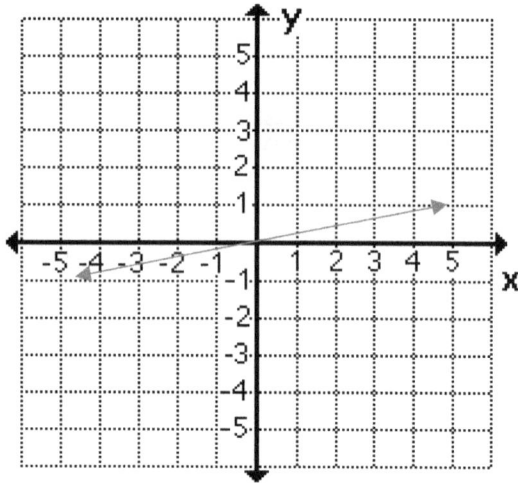

B.

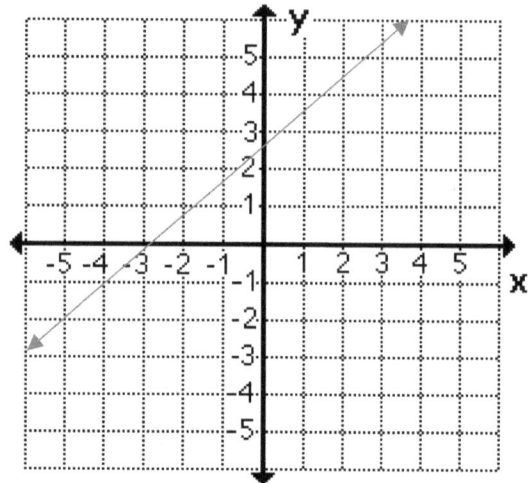

C.

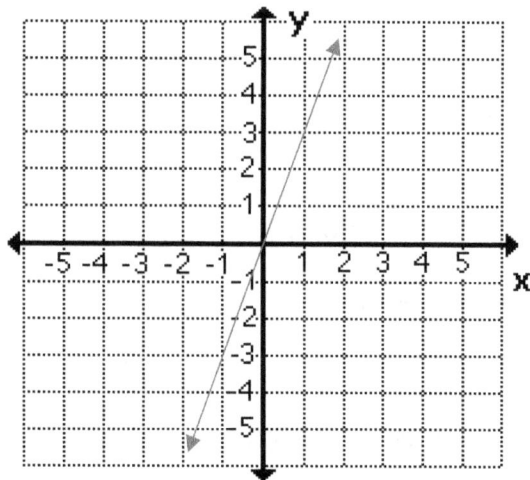

D.

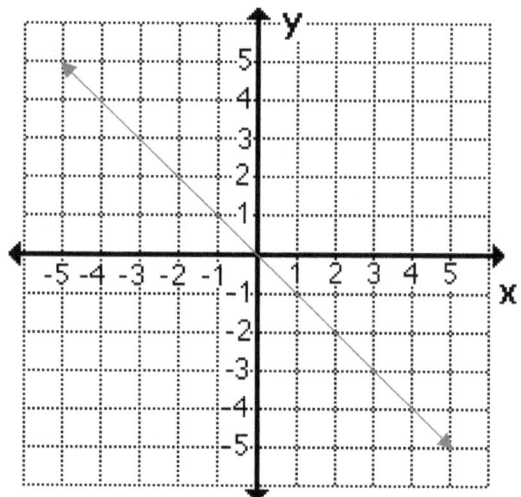

11) Three years ago, Amy was four times as old as Mike was. If Mike is 10 years old now, how old is Amy?

A. 31

C. 13

B. 28

D. 32

12) What is the value of $|-101 + 97| - |11(-2)|$?

    A. -18 C. -26

    B. -8 D. $+26$

13) What is the solution of the following system of equations?

$$\begin{cases} \dfrac{-x}{4} + \dfrac{y}{8} = -1 \\ \dfrac{4y}{5} + 2x = \dfrac{4}{5} \end{cases}$$

    A. $x = 2$, $y = -4$ C. $x = -4$, $y = -2$

    B. $x = 2$, $y = 6$ D. $x = -4$, $y = 2$

14) If a gas tank can hold 65 gallons, how many gallons does it contain when it is $\frac{3}{5}$ full?

    A. 39.2   C. 31

    B. 36   D. 39

15) What is the length of BC in the following figure if AB = 12, DF = 16 and BD = 49?

    A. 21

    B. 19

    C. 20

    D. 24

16) In the xy-plane, the point (-3,2) and (-2,6) are online A. Which of the following equations of lines is parallel to line A?

A. $y - x = 4$

B. $y = \frac{x}{4} + \frac{5}{3}$

C. $y = 3x + 4$

D. $5y - 20x = 5$

STOP

This is the end of Unit 1

Unit 2

Calculators are permitted for unit 2 of the test.

Read each question. Then mark your answers in your answer sheet.

If you have time, review your answers, and only answer questions you did not

answer in the unit.

Time for Unit 2: 80 Minutes

17) In the rectangle below if $y > 6$ cm and the area of rectangle is 42 cm^2 and the perimeter of the rectangle is 26 cm, what is the value of x and y respectively?

A. 7, 6

B. 6, 7

C. 2, 21

D. 3, 14

x ▭ y

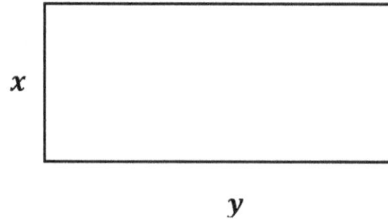

18) A football team had $26,000 to spend on supplies. The team spent $19,000 on new balls. New sport shoes cost $120 each. Which of the following inequalities represent the number of new shoes the team can purchase?

A. $120x + 19,000 \leq 26,000$

B. $120x + 19,000 \leq 26,000$

C. $19,000 + 120x \geq 26,000$

D. $19,000 + 120x \geq 26,000$

19) A $120 shirt now selling for $90 is discounted by what percent?

A. 30 %

C. 35 %

B. 20 %

D. 25 %

20) How much interest is earned on a principal of $5,000 invested at an interest rate of 3% for seven years?

A. $2,050

C. $2,000

B. $1,050

D. $1,000

21) A swimming pool holds 3,000 cubic feet of water. The swimming pool is 25 feet long and 20 feet wide. How deep is the swimming pool?

Write your answer in the box below.

```
[                    ]
```

22) The price of a car was $40,000 in 2014, $32,000 in 2015 and $25,600 in 2016. What is the rate of depreciation of the price of car per year?

A. 15 %

B. 25 %

C. 20 %

D. 30 %

23) The Jackson Library is ordering some bookshelves. If x is the number of bookshelves the library wants to order, which each cost $20 and there is a one-time delivery charge of $240, which of the following represents the total cost, in dollar, per bookshelf?

A. $20x + 240$

B. $20 + 240x$

C. $\dfrac{20x+240}{20}$

D. $\dfrac{20x+240}{x}$

24) The following table represents the value of x and function $f(x)$. Which of the following could be the equation of the function $f(x)$?

A. $f(x) = x^2 + x$

B. $f(x) = x^2 + 3$

C. $f(x) = \sqrt{3x + 4}$

D. $f(x) = 3\sqrt{x} + 1$

x	f(x)
1	4
4	7
9	10
16	13

25) The circle graph below shows all Mr. Wilson's expenses for last month. If he spent $940 on his car, how much did he spend for his rent?

A. $1,128

B. $1,235

C. $1,480

D. $1,210

Mr. Wilson's Monthly Expenses

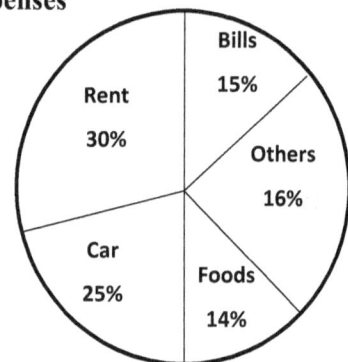

Rent 30%
Bills 15%
Others 16%
Foods 14%
Car 25%

26) The sum of seven different negative integers is -84. If the smallest of these integers is -15, what is the largest possible value of one of the other five integers?

A. -8

B. -12

C. -9

D. -7

27) In the following figure, point M lies on the line A, what is the value of y if $x = 25$?

A. 26

B. 24

C. 36

D. 25

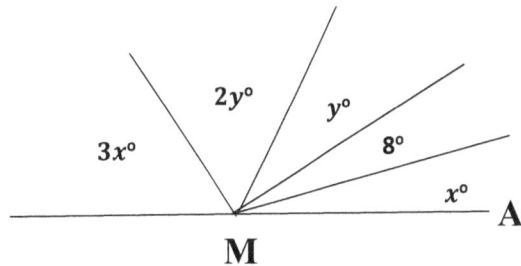

28) What is the smallest integer whose square root is greater than 10?

A. 120

C. 121

B. 110

D. 125

29) What is the area of the shaded region if the diameter of the bigger circle is 32 inches, and the diameter of the smaller circle is 26 inches?

A. $13\,\pi$

B. $16\,\pi$

C. $87\,\pi$

D. $78\,\pi$

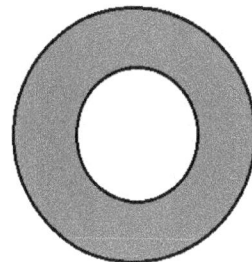

30) What is the sum of $\sqrt{x + 17}$ and $\sqrt{x} - 10$ when $\sqrt{x} = -8$?

A. -7

C. 7

B. -4

D. 4

STOP

This is the end of Unit 2

Unit 3

Calculators are permitted for unit 3 of the test.

Read each question. Then mark your answers in your answer sheet.

If you have time, review your answers, and only answer questions you did not answer in the unit.

Time for Unit 3: 80 Minutes

31) What is the area of an isosceles right triangle that has one leg that measures 10 cm?

Write your answer in the box below.

$$\boxed{}$$

32) If x is directly proportional to the square of y, and $y = 3$ when $x = 81$, then

if $x = 225$ $y = ?$

A. 4   C. 9

B. 5   D. 25

33) What is the value of x in the following figure?

A. 45

B. 44

C. 64

D. 72

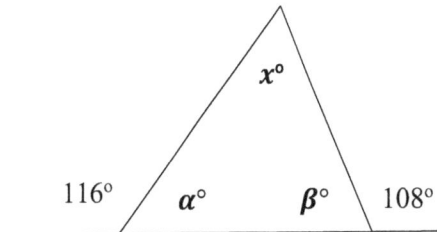

34) Jack earns $800 for his first 50 hours of work in a week and is then paid 1.5

times his regular hourly rate for any additional hours. This week, Jack needs

$1,136 to pay his rent, bills, and other expenses. How many hours must he work

to make enough money in this week?

A. 64   C. 14

B. 60   D. 16

Questions 35, 36 and 37 are based on the following data

Types of air pollutions in 10 cities of a country

Type of Pollution	Number of Cities									
A										
B										
C										
D										
E										
	1	2	3	4	5	6	7	8	9	10

35) If a is the mean (average) of the number of cities in each pollution type category, b is the mode, and c is the median of the number of cities in each pollution type category, then which of the following must be true?

A. $c < a < b$ C. $a = c$

B. $b < a < c$ D. $b < c = a$

36) How many cities should be added to type of pollutions C until the ratio of cities in type of pollution C to cities in type of pollution A will be 0.750?

A. 2 C. 4

B. 3 D. 5

37) What percent of cities are in the type of pollution B, C, and E respectively?

A. 30%, 50%, 60% C. 1.20%, 1.40%, 1.50%

B. 1.30%, 1.50%, 1.60% D. 20%, 40%, 50%

38) In the following right triangle, if the sides AB and AC become quadruple longer, what will be the ratio of the perimeter of the triangle to its area?

A. $\frac{4}{3}$

B. 1

C. $\frac{1}{4}$

D. 4

B

6 cm

A   8 cm   C

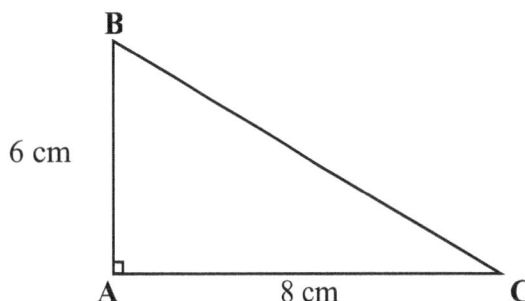

39) The capacity of a red box is 40% bigger than the capacity of a blue box. If the red box can hold 84 equal sized books, how many of the same books can the blue box hold?

A. 45

C. 55

B. 60

D. 50

40) A taxi driver earns $18 per hour work. If he works 14 hours a day, and he uses 4-liters Petrol in 2 hours with price $3 for 1-liter. How much money does he earn in one day?

A. $168

C. $158

B. $186

D. $165

Practice Test 1

This is the End of this Section.

Partnership for Assessment of Readiness for College and Careers

PARCC Practice Test 2

Mathematics

GRADE 8

Unit 1

Calculators are NOT permitted for unit 1 of the test.

Read each question. Then mark your answers in your answer sheet.

If you have time, review your answers, and only answer questions you did not answer in the unit.

Time for Unit 1: 80 Minutes

Administered Month Year

1) The area of a circle is $16\,\pi$. Which of the following can be the circumference of the circle?

    A. $4\,\pi$ C. $8\,\pi$

    B. $9\,\pi$ D. $7\,\pi$

2) You can buy 16 cans of green beans at a supermarket for $4.20. How much does it cost to buy 56 cans of green beans?

    A. $17.88   C. $14.70

    B. $28.90   D. $216.8

3) Which of the following is the solution of the following inequality?

$$4x - 4 > 12x - 8.5 - 3.5\,x$$

    A. $x < 2$ C. $x < 1$

    B. $x > 2$ D. $x > 1$

4) What is the perimeter of a square that has an area of 268.96 feet?

Write your answer in the box below.

$$\boxed{}$$

5) A tree 18 feet tall casts a shadow 54 feet long. Jack is 8 feet tall. How long is Jack's shadow?

    A. 24 ft   C. 70 ft

    B. 20.25 ft   D. 36 ft

6) The price of a laptop is decreased by 25% to $337.5. What is its original price?

A. 450

B. 275

C. 675

D. 540

7) A container holds 6.3 gallons of water when it is $\frac{7}{38}$ full. How many gallons of water does the container hold when it's full?

A. 38.8

B. 34.2

C. 24.5

D. 19.5

8) If $(7^a)^b = 2,401$, then what is the value of $a \times b$?

A. 12

B. 7

C. 6

D. 4

9) A bag contains 12 balls: two green, five black, three blue, one red and one white. If 11 balls are removed from the bag at random, what is the probability that a white ball has been removed?

A. $\frac{1}{12}$

B. $\frac{11}{12}$

C. $\frac{1}{11}$

D. $\frac{10}{11}$

10) What is the x-intercept of the line with equation $6x + 8y = 54$?

A. -9

B. $+8$

C. $+9$

D. $-\frac{8}{54}$

11) Which of the following expressions is equivalent to $7y(x - y)$?

    A. $7yx^2 - 7xy$ C. $-7y^2 + 7xy$

    B. $7x^2 - 7y^2$ D. $7xy^2 - 7x^2$

12) A soccer team played 140 games and won 45 percent of them. How many games did the team win?

    A. 26   C. 63

    B. 86   D. 39

13) The equation of a line is given as: $y = 3x - 2$. Which of the following points does not lie on the line?

    A. $(-1, -5)$ C. $(2, 4)$

    B. $(0, -2)$ D. $(1, -2)$

14) The perimeter of the trapezoid below is 40 cm. What is its area?

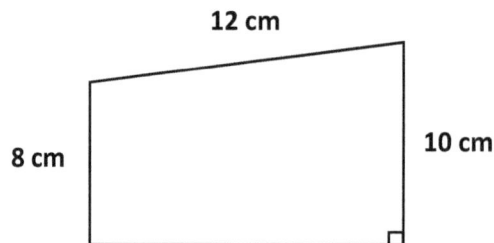

Write your answer in the box below.

    ┌─────────────┐
    │   │
    └─────────────┘

15) Which graph does not represent y as a function of x?

A.

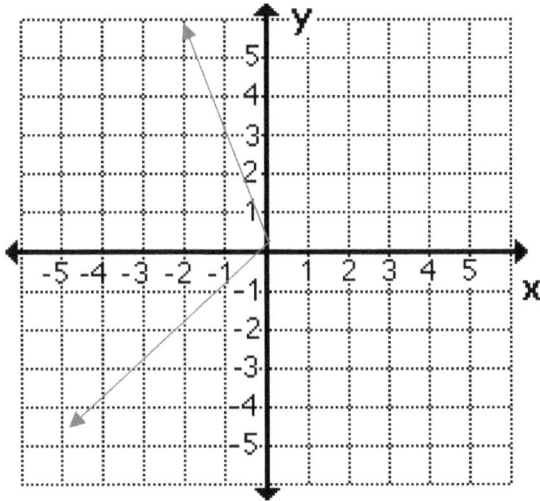

B.

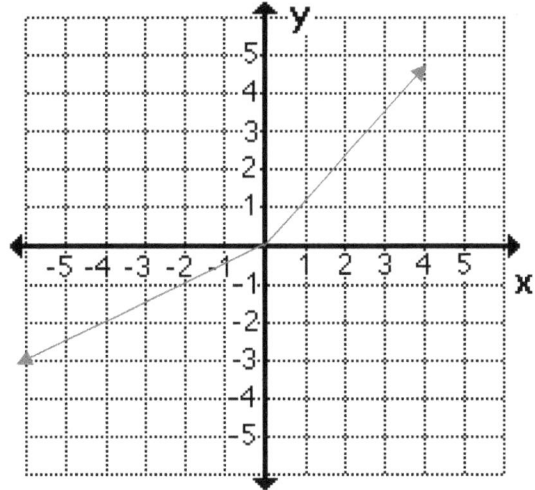

D.

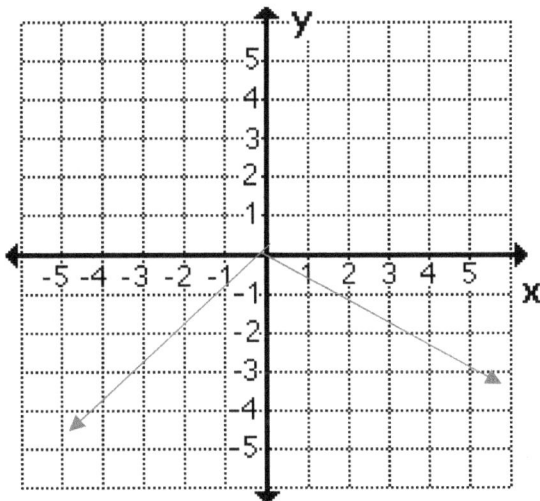

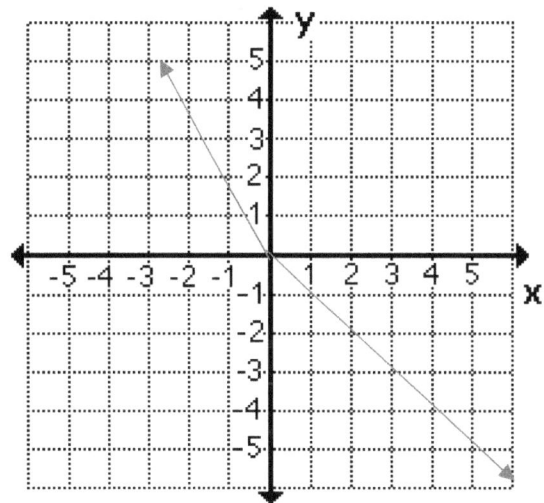

16) Which of the following is equivalent to $-27 < -10x + 3 < 3$?

A. $-3 < x < 0$

C. $3 < x < 0$

B. $0 < x < 3$

D. $-3 < x < -3$

17) A bank is offering 1.75% simple interest on a savings account. If you deposit $15,000, how much interest will you earn in four years?

A. $2,820   C. $1,280

B. $940   D. $1,050

STOP

This is the end of Unit 1

Unit 2

Calculators are permitted for unit 2 of the test.

Read each question. Then mark your answers in your answer sheet.

If you have time, review your answers, and only answer questions you did not answer in the unit.

Time for Unit 2: 80 Minutes

18) Joe scored 24 out of 32 marks in Algebra, 38 out of 42 marks in science and 25 out of 30 marks in mathematics. In which subject his percentage of marks is best?

A. Algebra

C. Mathematics

B. Science

D. Algebra and Science

19) What is the volume of the following triangular prism?

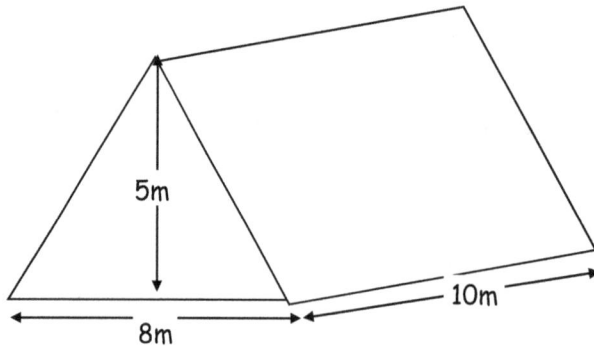

Write your answer in the box below.

[   ]

20) The marked price of a computer is E Euro. Its price decreased by 10% in March and later increased by 7 % in April. What is the final price of the computer in E Euro?

A. 0.903 E

C. 0.921 E

B. 0.991E

D. 0.963 E

21) Find the slope–intercept form of the graph $4x - 6y = -14$

A. $y = 3x + \dfrac{2}{7}$

C. $y = \dfrac{2}{3}x + 2\dfrac{1}{3}$

B. $y = \dfrac{2}{3}x + \dfrac{2}{7}$

D. $y = \dfrac{2}{3}x - 2\dfrac{1}{3}$

22) Triangle ABC is graphed on a coordinate grid with vertices at A $(6, 2)$, B $(-1, 3)$ and C $(4, 7)$. Triangle ABC is reflected over x axes to create triangle A' B' C'. Which order pair represents the coordinate of C'?

A. (7.4)

C. $(4. -7)$

B. $(-4. -7)$

D. $(-4. 7)$

23) Which of the following point is the solution of the system of equations?

$$\begin{cases} 4x + 3y = -5 \\ 3x + 2y = -3 \end{cases}$$

A. $(-3, 3)$

C. $(-1, 3)$

B. $(1, -3)$

D. $(-1, -3)$

24) A shirt costing $420 is discounted 15%. After a month, the shirt is discounted another 8%. Which of the following expressions can be used to find the selling price of the shirt?

A. $(420)(0.15)(0.8)$

B. $(420)(0.85)(0.92)$

C. $(420)(0.15)-(420)(0.80)$

D. $(420)(0.15)-(420)(0.92)$

25) What is the distance between the points $(1, 3)$ and $(-1, 3)$?

    A. 2

    B. 3

    C. 6

    D. 8

Questions 26 and 27 are based on the following data

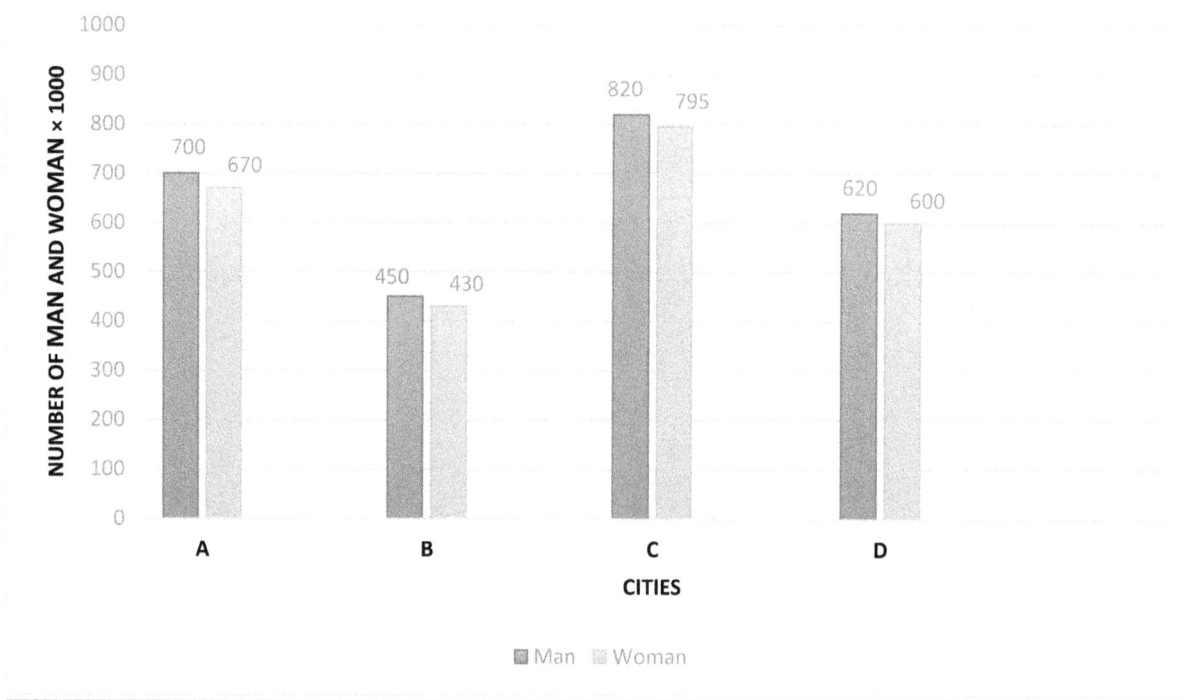

26) What's the ratio of percentage of women in city C to percentage of men in city B?

    A. 0.902   C. 0.853

    B. 0.921   D. 1. 109

27) What's the minimum ratio of woman to man in the four cities?

A. 0.95   C. 0.98

B. 0.97   D. 0.93

28) Line m passes through the point $(2, -3)$. Which of the following **CANNOT** be

the equation of line m?

A. $y = -3$ C. $y = 1 - x$

B. $y = -2x + 3$ D. $y = x - 5$

STOP
This is the end of Unit 2

Unit 3

Calculators are permitted for unit 3 of the test.

Read each question. Then mark your answers in your answer sheet.

If you have time, review your answers, and only answer questions you did not answer in the unit.

Time for Unit 3: 80 Minutes

29) The sum of four numbers is 58. If another number is added to these four numbers, the average of the four numbers is 18. What is the fifth number?

A. 28

C. 22.50

B. 20

D. 14

30) David owed $11,606. After making 41 payments of $198 each, how much did he have left to pay?

A. $3,488

C. $2,360

B. $3,680

D. $2,100

31) Which of the following lines is parallel to the graph of $y = 2x - 5$?

A. $3x - 5y = 4$

C. $4x - 2y = -6$

B. $2x - 7y = 0$

D. $2x + 4y = 6$

32) The average of three consecutive numbers is 20.5. What is the smallest number?

A. 14.5

C. 17.5

B. 19.5

D. 30

33) The price of a laptop is decreased by 25% to $450. What is its original price?

A. 750

C. 600

B. 650

D. 550

34) The average weight of 18 girls in a class is 35 kg and the average weight of 22 boys in the same class is 45 kg. What is the average weight of all the 40 students in that class?

A. 49 Kg

C. 41 Kg

B. 49.50 Kg

D. 40.50 Kg

35) Liam's average (arithmetic mean) on two mathematics tests is 80. What should Liam's score be on the next test to have an overall of 86 for all the tests?

A. 94

C. 99

B. 98

D. 93

36) Which of the following equations has a graph that is a straight line?

A. $3y = x^2 - 8$

C. $x - y = 8$

B. $y^2 - 2x^2 = 8$

D. $x + xy = 8$

37) An angle is equal to one seventh of its supplement. What is the measure of that angle?

A. 36

C. 32.5

B. 45.5

D. 22.5

38) Two seventh of 42 is equal to $\frac{3}{4}$ of what number?

A. 36

C. 16

B. 28

D. 46

39) A card is drawn at random from a standard 52–card deck, what is the probability that the card is of Spades? (The deck includes 13 of each suit clubs, diamonds, hearts, and spades)

A. $\frac{5}{52}$

C. $\frac{1}{2}$

B. $\frac{1}{4}$

D. $\frac{3}{13}$

40) The score of Zoe was one fourth of Emma and the score of Harper was thrice that of Emma. If the score of Harper was 96, what is the score of Zoe?

A. 16

C. 45

B. 26

D. 38

Practice Test 2

This is the End of this Section.

Chapter 12 : Answers and Explanations
PARCC Practice Tests

Answer Key

✳ Now, it's time to review your results to see where you went wrong and what areas you need to improve!

PARCC - Mathematics

Practice Test - 1							Practice Test - 2						
1	A	16	D	31	50		1	C	16	B	31	C
2	B	17	B	32	B		2	C	17	D	32	B
3	C	18	B	33	B		3	C	18	B	33	C
4	C	19	D	34	A		4	65.6	19	200	34	D
5	B	20	B	35	A		5	A	20	D	35	B
6	D	21	6	36	A		6	A	21	C	36	C
7	C	22	C	37	D		7	B	22	C	37	D
8	260	23	D	38	C		8	D	23	B	38	C
9	B	24	D	39	B		9	B	24	B	39	B
10	B	25	A	40	A		10	C	25	A	40	A
11	A	26	C				11	C	26	C		
12	A	27	B				12	C	27	B		
13	A	28	B				13	D	28	C		
14	D	29	C				14	90	29	D		
15	A	30	C				15	A	30	A		

Practice Test 1

Answers and Explanations

1) Answer: A

Use Pythagorean Theorem: $a^2 + b^2 = c^2$

$33^2 + 44^2 = C^2 \Rightarrow 1{,}089 + 1{,}936 = C^2 \Rightarrow 3{,}025 = c^2 \Rightarrow c = 55$

2) Answer: B

Let x be the number. Write the equation and solve for x. $\frac{(90-x)}{x} = 8$ (Cross multiply)

$(90 - x) = 8x$, then add x both sides. $90 = 9x$, now divide both sides by 9. $\Rightarrow x = 10$.

3) Answer: C

Use simple interest formula: $I = prt$; (I=interest, p=principal, r=rate, t=time)

$I = (8{,}000)(0.0375)(6) = 1{,}800$

4) Answer: C

Let x be the number of soft drinks for 158 guests. Write the proportion and solve for x.

$\frac{30 \text{ soft drinks}}{15 \text{guests}} = \frac{x}{158 \text{ guests}} \Rightarrow x = \frac{158 \times 30}{15} \Rightarrow x = 316$

5) Answer: B

12% of the volume of the solution is alcohol. Let x be the volume of the solution. Then: 12% of $x = 34.8$ ml

$0.12\,x = 34.8 \Rightarrow \frac{12x}{100} = \frac{348}{10}$ cross multiply; $120x = 34{,}800 \Rightarrow$ (devide by 120) $x = 290$

6) Answer: D

Use the area of rectangle formula (s = a × b).

To find area of the shaded region subtract smaller rectangle from bigger rectangle.

$S_1 - S_2 = (8\,ft \times 14ft) - (5\,ft \times 7ft) \Rightarrow S_1 - S_2 = 112 - 35 = 77ft.$

7) Answer: C

The weight of 46.4 meters of this rope is: $46.4 \times 300g = 13{,}920g$

1 kg = 1,000 g, therefore, $13{,}920\,g \div 1000 = 13.92$ kg

8) Answer: 260

The ratio of boy to girls is 5:7. Therefore, there are 5 boys out of 12 students. To find the answer, first divide the total number of students by 12, then multiply the result by 5.

$624 \div 12 = 52 \Rightarrow 52 \times 5 = 260$

9) Answer: B

the population is increased by 10% and 25%. 10% increase changes the population to 110% of original population.

For the second increase, multiply the result by 125%. $(1.1) \times (1.25) = 1.375 = 137.5\%$

37.5 percent of the population is increased after two years.

10) Answer: B

A linear equation is a relationship between two variables, x and y, that can be put in the form $y = mx + b$.

A non-proportional linear relationship takes on the form $y = mx + b$, where $b \neq 0$ and its graph is a line that does not cross through the origin.

11) Answer: A

Three years ago, Amy was four times as old as Mike. Mike is 10 years now. Therefore, 3 years ago Mike was 7 years.

Three years ago, Amy was: $A = 4 \times 7 = 28$; Now Amy is 31 years old: $28 + 3 = 31$

12) Answer: A

$|-101 + 97| - |11(-2)| = |-4| - |-22| = 4 - 22 = -18$

13) Answer: D

$\begin{cases} \frac{-x}{4} + \frac{y}{8} = -1 \\ \frac{4y}{5} + 2x = \frac{4}{5} \end{cases} \rightarrow$ Multiply the top equation by 8. Then, $\begin{cases} -2x + y = -8 \\ \frac{4y}{5} + 2x = \frac{4}{5} \end{cases} \rightarrow$ Add two equations.

$\frac{9}{5}y = \frac{-36}{5} \rightarrow y = -4$, plug in the value of y into the first equation $\rightarrow x = 2$

14) Answer: D

$\frac{3}{5} \times 65 = \frac{195}{5} = 39$

15) Answer: A

Two triangles ΔABC and ΔCDF are similar. Then:

$\frac{AB}{DF} = \frac{BC}{CD} \rightarrow \frac{12}{16} = \frac{3}{4} = \frac{x}{49-x} \rightarrow 147 - 3x = 4x \rightarrow 7x = 147 \rightarrow x = 21$

16) Answer: D

The slop of line A is: $m = \frac{y_2 - y_1}{x_2 - x_1} = \frac{6-2}{-2-(-3)} = 4$

Parallel lines have the same slope.

Choice D $(5y - 20x = 5 \Rightarrow y = 4x + 1)$ has slope of 4.

17) Answer: B

The perimeter of the rectangle is: $2x + 2y = 26 \rightarrow x + y = 13 \rightarrow x = 13 - y$

The area of the rectangle is: $x \times y = 42 \rightarrow (13 - y)(y) = 42 \rightarrow y^2 - 13y + 42 = 0$

Solve the quadratic equation by factoring method.

$(y - 6)(y - 7) = 0 \rightarrow y = 6$ (Unacceptable, because y must be greater than 6) or $y = 7$; If $y = 7 \rightarrow x = 13 - y \rightarrow x = 13 - 7 \rightarrow x = 6$

18) Answer: B

Let x be the number of new shoes the team can purchase. Therefore, the team can purchase $120\,x$.

The team had \$26,000 and spent \$19,000. Now the team can spend on new shoes \$7,000 at most.

Now, write the inequality: $120x + 19,000 \leq 26,000$

19) Answer: D

Use the formula for Percent of Change

$\frac{\text{New Value} - \text{Old Value}}{\text{Old Value}} \times 100\% \Rightarrow \frac{90 - 120}{120} \times 100\% = -25\%$

(Negative sign here means that the new price is less than old price).

20) Answer: B

Use simple interest formula: $I = prt$ (I = interest, p = principal, r = rate, t = time)

I = (5,000) (0.03) (7) = 1,050

21) Answer: 6

Use formula of rectangle prism volume.

V = (length) (width) (height) $\Rightarrow 3,000 = (25) (20)$ (height) $\Rightarrow$

height = $3,000 \div 500 = 6$

22) Answer: C

Use this formula: Percent of Change $= \frac{\text{New Value} - \text{Old Value}}{\text{Old Value}} \times 100\%$

$\frac{32,000 - 40,000}{40,000} \times 100\% = -20\%$ and $\frac{25,600 - 32,000}{32,000} \times 100\% = -20\%$

23) Answer: D

The amount of money for x bookshelf is: $20x$

Then, the total cost of all bookshelves is equal to: $20x + 240$

The total cost, in dollar, per bookshelf is: $\frac{Total\ cost}{number\ of\ items} = \frac{20x + 240}{x}$

24) Answer: D

A. $f(x) = x^2 + x$ if $x = 4 \rightarrow f(4) = (4)^2 + 4 = 20 \neq 7$

B. $f(x) = x^2 + 3$ if $x = 4 \rightarrow f(4) = (4)^2 + 3 = 19 \neq 7$

C. $f(x) = \sqrt{3x + 4}$ if $x = 4 \rightarrow f(4) = \sqrt{3(4) + 4} = \sqrt{16} = 4 \neq 7$

D. $f(x) = 3\sqrt{x} + 1$ if $x = 4 \rightarrow f(4) = 3\sqrt{4} + 1 = 7 = 7$

25) Answer: A

Let x be all expenses, then

$$\frac{25}{100}x = \$940 \rightarrow x = \frac{100 \times \$940}{25} = \$3,760$$

He spent for his rent: $\frac{30}{100} \times \$3,760 = \$1,128$

26) Answer: C

The smallest number is -15. To find the largest possible value of one of the other seven integers, we need to choose the smallest possible integers for six of them. Let x be the largest number. Then:

$$-84 = (-15) + (-14) + (-13) + (-12) + (-11) + (-10) + x \rightarrow -84 = -75 + x$$

$$\rightarrow x = -84 + 75 = -9$$

27) Answer: B

The angles on a straight line add up to 180 degrees. Then:

$x + 8 + y + 2y + 3x = 180$

Then, $4x + 3y = 180 - 8 \rightarrow 4(25) + 3y = 172 \rightarrow 3y = 172 - 100 = 72 \rightarrow 3y = 72$

$\rightarrow y = 24$

28) Answer: B

Square root of 120 is $\sqrt{120} = \sqrt{100 + 20} > \sqrt{100} = 10$

Square root of 110 is $\sqrt{110} = \sqrt{100 + 10} > \sqrt{100} = 10$

Square root of 121 is $\sqrt{121} = 11 > \sqrt{100} = 10$

Square root of 125 is $\sqrt{125} = \sqrt{121 + 4} > \sqrt{100} = 10$

29) Answer: C

To find the area of the shaded region subtract smaller circle from bigger circle.

$S_{\text{bigger}} - S_{\text{smaller}} = \pi \, (r_{\text{bigger}})^2 - \pi \, (r_{\text{smaller}})^2 \Rightarrow S_{\text{bigger}} - S_{\text{smaller}} = \pi \, (16)^2 - \pi \, (13)^2$

$\Rightarrow 256\,\pi - 169\pi = 87\,\pi$

30) Answer: C

$\sqrt{x} = -8 \rightarrow x = 64$

then; $\sqrt{x} - 10 = \sqrt{64} - 10 = 8 - 10 = -2$ and $\sqrt{x + 17} = \sqrt{64 + 17} = \sqrt{81} = 9$

Then: $\left(\sqrt{x + 17}\right) + \left(\sqrt{x} - 10\right) = 9 + (-2) = 7$

31) Answer: 50.

$a = 10 \Rightarrow$ area of the triangle is $= \frac{1}{2}(10 \times 10) = \frac{100}{2} = 50 \ cm^2$

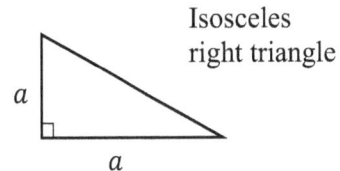

Isosceles right triangle

32) Answer: B

x is directly proportional to the square of y. Then: $x = cy^2$

$81 = c(3)^2 \rightarrow 81 = 9c \rightarrow c = \frac{81}{9} = 9$

The relationship between x and y is: $x = 9y^2$, $x = 225$

$225 = 9y^2 \rightarrow y^2 = \frac{225}{9} = 25 \rightarrow y = 5$

33) Answer: B

$\alpha = 180° - 116° = 64°$

$\beta = 180° - 108° = 72°$

$x + \alpha + \beta = 180° \rightarrow x = 180° - 64° - 72° = 44°$

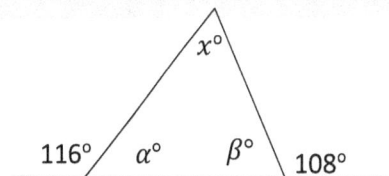

34) Answer: A

The amount of money that jack earns for one hour: $\frac{\$800}{50} = \$16$

Number of additional hours that he works to make enough money is: $\frac{\$1,136 - \$800}{1.5 \times \$16} = 14$

Number of total hours is: $50 + 14 = 64$

35) Answer: A

Let's find the mean (average), mode and median of the number of cities for each type of pollution.

Number of cities for each type of pollution: 8, 2, 4, 8, 5

$average \ (mean) = \frac{sum \ of \ terms}{number \ of \ terms} = \frac{8+2+4+5+8}{5} = \frac{27}{5} = 5.4$

Median is the number in the middle. To find median, first list numbers in order from smallest to largest: 2, 4, 5, 8, 8; and Median of the data is 5.

Mode is the number which appears most often in a set of numbers. Therefore, mode in the set of numbers is 8.

Median < Mean<Mode: then, c < a < b

36) Answer: A

Let the number of cities should be added to type of pollutions C be x. Then:

$\frac{x+4}{8} = 0.750 \rightarrow x + 4 = 8 \times (0.750) \rightarrow x + 4 = 6 \rightarrow x = 2$

37) Answer: D

Percent of cities in the type of pollution B: $\frac{2}{10} \times 100 = 20\%$

Percent of cities in the type of pollution C: $\frac{4}{10} \times 100 = 40\%$

Percent of cities in the type of pollution E: $\frac{5}{10} \times 100 = 50\%$

38) Answer: C

$AB = 6$, And $AC = 8$

$BC = \sqrt{6^2 + 8^2} = \sqrt{36 + 64} = \sqrt{100} = 10$

Perimeter $= 6 + 8 + 10 = 24$; Area $= \frac{6 \times 8}{2} = 24$

In this case, the ratio of the perimeter of the triangle to its area is: $\frac{24}{24} = 1$

If the sides AB and AC become triple longer, then:

$AB = 24$, And $AC = 32$; $BC = \sqrt{24^2 + 32^2} = \sqrt{576 + 1,024} = \sqrt{1,600} = 40$

Perimeter $= 24 + 32 + 40 = 96$; Area $= \frac{32 \times 24}{2} = 384$

In this case the ratio of the perimeter of the triangle to its area is: $\frac{96}{384} = \frac{1}{4}$

39) Answer: B

The capacity of a red box is 40% bigger than the capacity of a blue box and it can hold 84 books. Therefore, we want to find a number that 20% bigger than that number is 30. Let x be that number. Then: $1.40 \times x = 84$, Divide both sides of the equation by 1.4. Then: $x = \frac{84}{1.40} = 60$

40) Answer: A

$\$18 \times 14 = \$252$

Petrol use: $4 \div 2 = 2$, $14 \times 2 = 28$ liters; Petrol cost: $28 \times \$3 = \$84$

Money earned: $\$252 - \$84 = \$168$

Practice Test 2

Answers and Explanations

1) Answer: C

Use the formula for area of circles. Area $= \pi r^2 \Rightarrow 16\pi = \pi r^2 \Rightarrow 16 = r^2 \Rightarrow r = 4$

Radius of the circle is 4. Now, use the circumference formula:

Circumference $= 2\pi r = 2\pi (4) = 8\pi$

2) Answer: C

Let x be the number of cans. Write the proportion and solve for x.

$\frac{16 \text{ cans}}{\$ 4 \cdot 20} = \frac{56 \text{ cans}}{x} \Rightarrow x = \frac{56 \times 4 \cdot 20}{16} \Rightarrow x = \$ 14.70$

3) Answer: C

$4x - 4 > 12x - 8.5 - 3.5x \rightarrow$ Combine like terms:

$4x - 4 > 8.5x - 8.5 \rightarrow$ Subtract $4x$ from both sides: $-4 > 4.5x - 8.5$

Add 8.5 both sides of the inequality.

$4.5 > 4.5x$, Divide both sides by 4.5. $\Rightarrow \frac{4.5}{4.5} > x \rightarrow x < 1$

4) Answer: 65.6 feet.

Area of a square: $S = a^2 \Rightarrow 268.96 = a^2 \Rightarrow a = 16.4$

Perimeter of a square: $P = 4a \Rightarrow P = 4 \times 16.4 \Rightarrow P = 65.6$

5) Answer: A

Write the proportion and solve for the missing number.

$\frac{18}{54} = \frac{8}{x} \rightarrow 18x = 8 \times 54 = 432 \rightarrow 18x = 432 \rightarrow x = \frac{432}{18} = 24$

6) Answer: A

Let x be the original price.

If the price of a laptop is decreased by 25% to $337.5, then:

75 % of x $= 337.5 \Rightarrow 0.75 x = 337.5 \Rightarrow x = 337.5 \div 0.75 = 450$

7) Answer: B

let x be the number of gallons of water the container holds when it is full.

Then; $\frac{7}{38}x = 6.3 \rightarrow x = \frac{38 \times 6.3}{7} = 34.2$

8) Answer: D

$(7^a)^b = 2.401 \rightarrow 7^{ab} = 2.401 \rightarrow 2.401 = 7^4 \rightarrow 7^{ab} = 7^4 \rightarrow ab = 4$

9) Answer: B

If 11 balls are removed from the bag at random, there will be one ball in the bag.

The probability of choosing a white ball is 1 out of 12. Therefore, the probability of not choosing a white ball is 11 out of 12 and the probability of having not a white ball after removing 11 balls is the same.

10) Answer: C

The value of y in the x-intercept of a line is zero. Then:

$y = 0 \rightarrow -6x + 8(0) = 54 \rightarrow 6x = 54 \rightarrow x = \frac{54}{6} = 9$; then, x-intercept of the line is 9

11) Answer: C

Use distributive property: $7y(x - y) = 7y(x) + 7y(-y) = 7yx - 7y^2$

12) Answer: C

$140 \times \frac{45}{100} = 63.$

13) Answer: D

$y = 3x - 2$

$(-1. -5) \Rightarrow -5 = 3(-1) - 2 \Rightarrow -5 = -5$

$(0. -2) \Rightarrow -2 = 3(0) - 2 \Rightarrow -2 = -2$

$(2. 4) \Rightarrow 4 = 3(2) - 2 \Rightarrow 4 = 4$

$(1. -2) \Rightarrow -2 = 3(1) - 2 \Rightarrow -2 \neq 1$

14) Answer: 90.

The perimeter of the trapezoid is 40 cm.

Therefore, the missing side (height) is $= 40 - 10 - 8 - 12 = 10$

Area of a trapezoid: A $= \frac{1}{2}$ h (b$_1$ + b$_2$) $= \frac{1}{2}$ (10) (10 + 8) = 90

15) Answer: A

A graph represents y as a function of x if $x_1 = x_2 \rightarrow y_1 = y_2$

In choice A, for each x, we have two different values for y.

16) Answer: B

$-27 < -10x + 3 < 3 \rightarrow$ Subtract 3 to all sides.

$-27 - 3 < -10x + 3 - 3 < 3 - 3 \rightarrow -30 < -10x < 0 \rightarrow$ Divide all sides by $- 10$.

(Remember that when you divide all sides of an inequality by a negative number, the inequality sing will be swapped. $<$ becomes $>$): $\frac{-30}{-10} < \frac{-10x}{-10} < \frac{0}{-10} \Rightarrow 3 > x > 0$, or $0 < x < 3$

17) Answer: D

Use simple interest formula: $I = prt$(I = interest, p = principal, r = rate, t = time)

I = (15.000) (0.0175) (4) =1,050

18) Answer: B

Compare each mark:

In Algebra Joe scored 24 out of 32 in Algebra. It means Joe scored 75% of the total mark. $\frac{24}{32} =$

$\frac{x}{100} \Rightarrow x = 75\%$

Joe scored 38 out of 42 in science. It means Joe scored 90% of the total mark.

$\frac{38}{42} = \frac{x}{100} \Rightarrow x = 90\%$

Joe scored 25 out of 30 in mathematic that it means 83% of total mark.

$\frac{25}{30} = \frac{x}{100} \Rightarrow x = 83\%$

Therefore, his score in Scored a is higher than his other scores.

19) Answer: 200 m³.

Use the volume of the triangular prism formula.

V = $\frac{1}{2}$ (length) (base) (high) →V = $\frac{1}{2} \times 10 \times 8 \times 5 \Rightarrow$ V = 200 m³

20) Answer: D

To find the discount, multiply the number by (100% – rate of discount).

Therefore, for the first discount we get: $(100\% - 10\%)(E) = (0.90)E$

For increase of 7 %:

$(0.90) E \times (100\% + 7\%) = (0.90) (1.07) = 0.963E$.

21) Answer: C

$4x - 6y = -14 \Rightarrow -6y = -4x - 14 \Rightarrow y = \frac{-4}{-6}x - \frac{14}{-6} \Rightarrow y = \frac{2}{3}x + \frac{7}{3}$

$y = \frac{2}{3}x + 2\frac{1}{3}$

22) Answer: C

When a point is reflected over x axes, the (y) coordinate of that point changes to $(-y)$ while its x coordinate remains the same. C (4, 7) → C' (4, −7)

23) Answer: B

Solving Systems of Equations by Elimination

$$4x + 3y = -5$$
$$3x + 2y = -3$$
Multiply the first equation by 3, and second equation by -4, then add two equations.

$$\begin{array}{l} 3(4x + 3y = -5) \\ -4(3x + 2y = -3) \end{array} \Rightarrow \begin{array}{l} 12x + 9y = -15 \\ -12x - 8y = 12 \end{array} \Rightarrow y = -3.$$

$$3x + 2y = -3, 3x + 2(-3) = -3, \text{then: } x = 1, (1, -3)$$

24) Answer: B

To find the discount, multiply the number by (100% − rate of discount).

Therefore, for the first discount we get: (420) (100% − 15%) = (420) (0.85)

For the next 8 % discount: (420) (0.85) (0.92).

25) Answer: A

Use distance formula:

$$C = \sqrt{(x_A - x_B)^2 + (y - y_B)^2} \Rightarrow C = \sqrt{(1 - (-1))^2 + (3 - 3)^2}$$
$$C = \sqrt{(2)^2 + (0)^2} \Rightarrow C = \sqrt{4 + 0} \Rightarrow C = \sqrt{4} = 2$$

26) Answer: C

Percentage of women in city C $= \dfrac{795}{1,615} \times 100 = 49.22\%$

Percentage of men in city B $= \dfrac{450}{780} \times 100 = 57.69\%$

Percentage of men in city B to percentage of women in city C: $\dfrac{49.22}{57.69} = 0.853$

27) Answer: B

Ratio of women to men in city A: $\dfrac{670}{700} = 0.96$

Ratio of women to men in city B: $\dfrac{430}{450} = 0.96$

Ratio of women to men in city C: $\dfrac{795}{820} = 0.97$

Ratio of women to men in city D: $\dfrac{600}{620} = 0.97$

28) Answer: C

Solve for each equation: $(2, -3)$

$y = -3 \Rightarrow -3 = -3$

$y = -2x + 3 \Rightarrow -3 = (-2)(3) + 3 \Rightarrow -3 = -3$

$y = 1 - x \Rightarrow -3 = 1 - 2 \Rightarrow -1 \neq -3$

$y = x - 5 \Rightarrow -3 = 2 - 5 \Rightarrow -3 = -3$

29) Answer: D

$a + b + c + d = 58 \Rightarrow \frac{a+b+c+d+e}{4} = 18 \Rightarrow a + b + c + d + e = 72$

$\Rightarrow 58 + e = 72 \Rightarrow e = 72 - 58 = 14$

30) Answer: A

$41 \times \$198 = \$8,118$ Payable amount is: $\$11,606 - \$8,118 = \$3,488$

31) Answer: C

If two lines are parallel with each other, then the slope of the two lines is the same.

Then in line $y = 2x - 5$, the slope is equal to 2

And in the line $4x - 2y = -6 \Rightarrow y = 2x + 3$, the slope equal to 2

32) Answer: B

Let x be the smallest number. Then, these are the numbers: $x, x + 1, x + 2$

$\text{average} = \frac{\text{sum of terms}}{\text{number of terms}} \Rightarrow 20.5 = \frac{x+(x+1)+(x+2)}{3} \Rightarrow 20.5 = \frac{3x+3}{3} \Rightarrow 61.5 = 3x + 3 \Rightarrow 58.5 = 3x$

$\Rightarrow x = 19.5$

33) Answer: C

Let x be the original price.

If the price of a laptop is decreased by 25% to $450, then:

$75 \% \ of \ x = 450 \Rightarrow 0.75 \ x = 450 \Rightarrow x = 450 \div 0.75 = 600$

34) Answer: D

The sum of the weight of all girls is: $18 \times 35 = 630$ kg

The sum of the weight of all boys is: $22 \times 45 = 990$ kg

The sum of the weight of all students is: $630 + 990 = 1,620$ kg

$\text{average} = \frac{\text{sum of terms}}{\text{number of terms}}$; $\text{average} = \frac{1,620}{40} = 40.5$

35) Answer: B

$\frac{a + b}{2} = 80 \Rightarrow a + b = 160$

$\frac{a + b + c}{3} = 86 \Rightarrow a + b + c = 258$

$160 + c = 258 \Rightarrow c = 258 - 160 = 98$

36) Answer: C

$x - y = 8$ has a graph that is a straight line. All other options are not equations of straight lines.

37) Answer: D

The sum of supplement angles is 180. Let x be that angle. Therefore,

$x + 7x = 180$; $8x = 180$, divide both sides by 8: $x = 22.5$

38) Answer: C

Let x be the number. Write the equation and solve for x.

$\frac{2}{7} \times 42 = \frac{3}{4} \cdot x \Rightarrow \frac{2 \times 42}{7} = \frac{3x}{4}$, use cross multiplication to solve for x.

$4 \times 84 = 3x \times 7 \Rightarrow 336 = 21x \Rightarrow x = 16$

39) Answer: B

The probability of choosing a Spades is $\frac{13}{52} = \frac{1}{4}$

40) Answer: A

If the score of Harper was 96, therefore the score of Emma is 32. Since, the score of Zoe was one fourth of Emma, therefore, the score of Zoe is 8.

"End"

www.ingramcontent.com/pod-product-compliance
Lightning Source LLC
Chambersburg PA
CBHW080510090426
42734CB00015B/3019